ATLAS OF COLOSCOPY

Francesco Paolo Rossini, M.D.

Department of Gastroenterology,
Ospedale Maggiore
di S. Giovanni Battista e della Città di Torino,
Torino, Italy.

Atlas of coloscopy

Preface by Mario Banche, M.D.
Chief, Department of Gastroenterology,
Ospedale Maggiore
di S. Giovanni Battista e della Città di Torino,
Torino, Italy.

Translation by
V. Cameron-Curry, M. D.

Piccin Medical Books
Padova - Italy

Printed in Italy

To my wife and children

Contents

PART THREE

PART FOUR

Preface

by Mario Banche, M.D.

The neologism "coloscopy" (colonscopy, colonoscopy) is the latest addition to the vocabulary of endoscopy of the large intestine, just as the term "duodenoscopy" was in endoscopy of the upper digestive tract a few years ago. With the advent of the coloscope it is now possible to examine the large bowel more extensively and effectively than was formerly possible with the use of its rigid forerunner, the rectosigmoidoscope. The earliest rectoscopes, "endoscopes" introduced during the 19th century by Segales and Désormeaux (1826, 1853), were conceived for many uses—inspection of the urethra, bladder, uterus, rectum.

Successive improvements eventually led to the construction of an instrument designed exclusively for endoscopy of the rectum and sigmoid colon (Bensaude, 1907). Over the following years the rectoscope underwent no substantial change and thus an abundant and homogeneous literature accrued in which the available instruments and their use were described while the endoscopic pathologic nosography and corresponding endoscopic pictures of the rectum and sigmoid colon were codified, illustrated first by simple sketches and later by still and motion photography. The literature contains many excellent publications, some in monograph form.

No substantial advance in endoscopy of the large intestine, with the exception of the rectum and sigmoid colon, was made until about 1960, so much so that the current belief was that pathologic conditions of the large bowel, especially tumours, were confined largely to its end sections and thus exclusively concerned the field of rectosigmoidoscopy. However, after almost 60 years, the diagnostic gap has rapidly been bridged by coloscopy. This technique, without lessening the importance of rectosigmoidoscopy in the diagnosis of diseases of the rectum and sigmoid colon, has extended the possibility of endoscopic examination to the whole colon up to and even beyond the ileocecal valve. Coloscopy is now established as a method for the early detection of malignancies of the colon and for their effective prevention by endoscopic polypectomy.

Thanks to this new diagnostic technique it is now possible to detect the early stages and to establish the pathological anatomical background of a variety of often mild, occasional symptoms which can be investigated by no other instrumental, radiological or laboratory method. Minor, irregular and inconstant abnormalities of the bowel function, mild rectal bleeding, myxorrhoea, etc., symptoms often of no clinical significance, can now be investigated to rule out or confirm early granulomatous colitis, a mild recurrence of suprasigmoid haemorrhagic rectocolitis, single or multiple radiologically undemonstrable polyps, carcinoma; conversely, a doubtful radiological picture can be proved to be innocent.

A wealth of information obtained in vivo has thus increased our diagnostic possibilities in a variety of well-known disorders which may otherwise escape attention in their early stages or whose severity and extension may be underestimated.

The purpose of this book is to make available to the medical profession generally, and to students of gastroenterology specially, the most up-to-date information available on this new

endoscopic technique, its birth and rapid development, the instruments required, the technique, and its new uses.
Francesco Paolo Rossini was immediately fascinated by this new endoscopic technique. Ever since the necessary equipment was made available to him in our Department of Gastroenterology he has devoted his time and efforts mainly to this type of endoscopy and consequently to the investigation of diseases of the large intestine in which coloscopy represents an exceptional, but effective, method of examination. He has fast acquired great competence in this field and collected such a wealth of data that he has been persuaded to describe the results obtained in a great variety of cases in this monograph. A predominant place having been reserved for the illustrations, the book has taken on the form of an atlas.
After brief summary of the history of coloscopy, the equipment, preparation for coloscopy, method of examination, and the indications, contraindications and possible complications of coloscopy are described. The following chapters represent the most important part of the book and illustrate the endoscopic appearance of the normal colon and the endoscopic pictures found in aspecific and specific infectious and parasitic diseases, in cases of abnormal motor function (irritable colon) and structural abnormalities of the colon, with special emphasis on diverticulosis. The chapters on the two most common and characteristic inflammatory diseases of the colon, ulcerative colitis and granulomatous colitis, illustrate a great number of original and hitherto unpublished observations. The next chapters deal with tumours of the colon, from hyperplastic and adenomatous polyps to carcinoma. A separate chapter is devoted to the technique of peroperative coloscopy. It is mostly due to the pen of Alberto Montori, Professor of Surgery in the Surgical Clinic of the Rome University Medical School whose Chief is professor Giovanni Marcozzi.
Professor Montori was the first to develop this surgical endoscopic method. A short chapter on emergency coloscopy in rectal bleeding is followed by an important chapter on the technique of endoscopic polypectomy and the prevention of cancer of the large intestine which ends the Atlas This extremely interesting technique represents the latest advance in fibreoptic endoscopy, in a field in which the results obtained had previously been few. Having devoted much attention to this ancillary coloscopic technique from its inception, the author can now offer a complete and detailed documentation of the method, stressing its advantages and importance.
The facilities offered by the Publisher and his great experience have much lightened Dr. Rossini's task. The resulting Atlas is a work worthy of the consideration not only of endoscopists, but also of all gastroenterologists and of the medical profession generally.

Acknowledgments

I wish to express my deepest gratitude to my chief, professor Mario Banche, for his guidance and incitement and for the confidence and enthusiasm which he has aroused in me, without which this book would never have been written. My thanks to all those who have assisted me in bringing my task to an end. I am especially indebted to professor G. Filogamo, for his advice, to professor A. Montori, who wrote the chapter on peroperative coloscopy, to professor R. Cirillo, who did the radiological work, to G. Brunazzi, the art director who arranged the design of the atlas, to all the doctors who sent me clinical material, and to the nursing staff, G. Giannotti, A. Ottavis and L. Canazza, who patiently collaborated with me in carrying out the endoscopic examinations. My special thanks are due to the publisher, dr. M. Piccin, for accepting my book for publication and for the impeccable typographical result.

F.P.R.

Part one

Stages in the development of coloscopy

1795 Bozzini anorectoscope with candle lighting.
1826 Bozzini anorectoscope modified by Segales.
1865 Désormeaux rectosigmoidoscope with alcohol lamp.
1905 Tuttle electric proctoscope.
1907 Bensaude rectosigmoidoscope.
1957-8 Matsunaga sigmoidocamera, models 1, 2 & 3.
1961 Matsunaga cavocamera, models 1 & 2.
1962 Turrel fibreoptic sigmoidoscope.
1963 Sigmoidocamera introduced by the monorail method.
1964-6 Watanabe fibrecoloscope, types 1, 2 & 3.
Matsunaga fibrecoloscope, model 1.
1966 Watanabe type 4 fibrecoloscope with manoeuvrable tip.
1966-7 Total end-to-end transanal coloscopy (Provenzale, Camerada and Revignas).
Coloscopy by retrograde intubation (Torsoli, Arullani and Paoluzzi).
1968 Matsunaga fibrecoloscope, model 2.
1969 Matsunaga fibrecoloscope, model 3, with multiple direction tip angulation.
Watanabe fibrecoloscope, type 6.
1970 Watanabe fibrecoloscope, type 7, with multiple direction tip angulation.
Overholt fibrecoloscope with multiple direction tip angulation.

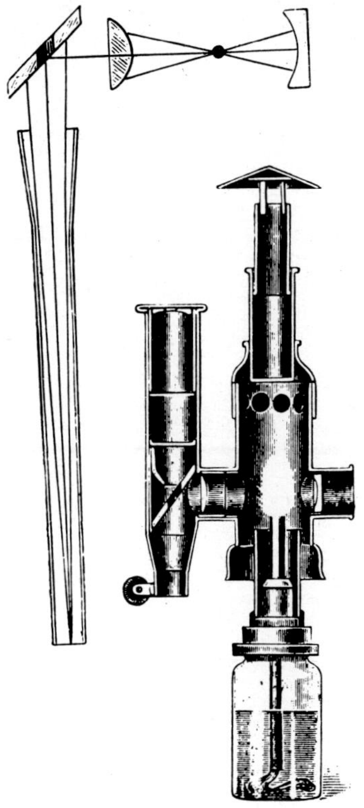

FIG. 1.
Endoscope de Désormeaux.
1865

Fig. 1. (after Moutier).

Further progress has led to the introduction of highly sophisticated intruments of great technical perfection such as the F9-A ACMI, FCS Machida and CF-LB2 Olympus coloscopes.
The acrobatic skill of our forerunners, however, has left an indelible mark in the history of coloscopy.

Preparation for coloscopy

bowel preparation and dietary measures

Satisfactory endoscopic examination of the colon requires that the large intestine be completely cleared of faeces.
The patient's collaboration is necessary, although the most important task is that of the nursing staff who must ensure that the endoscopist's instructions are carried out meticulously.
Endoscopy may be feasible in the presence of small amounts of soft faeces, enema residues, or mucus which can all be easily removed by aimed washout and suction through the coloscope.
Solid faeces cannot be removed in this way because the suction channel of the instrument measures only 2 mm in diameter.
Suffusion of blood in the mucosa may be caused by prolonged aspiration (Fig. 2).
Various agents have been used with the aim of cleansing the large intestine completely. Castor oil or magnesium sulphate (30 g daily for two days) associated with a residue-free diet can effectively clear the colon. However, it has often been noted that both castor oil and magnesium sulphate can cause mucosal changes in the large intestine (hyperaemia, oedema, congestion of the vascular networks) and especially in the last loop of the ileum where they may simulate aspecific or granulomatous inflammation.
Preparation of hospitalised patients for coloscopy:

— 1st day: the patient is allowed 200 g of meat, two portions of milk pudding, and unlimited amounts of water. One tablet of Senakot is given every 8 hours. A drop by drop proctoclysis of one litre of saline solution is given in the evening.
— 2nd day: the diet is the same as on the 1st day. One tablet of Senakot is given every 8 hours. A proctoclysis of one litre of water with a teaspoonful of granular Senakot is given in the morning and in the evening. 500 ml of 10% glucose solution containing B group vitamins, ascorbic acid, and 10 ml of 10% saline solution and 500 ml of 5% aminoacid solution are also infused by vein. The daily calorie intake is calculated as 1,000 calories.
— 3rd day: no food is given, but unlimited amounts of water are allowed. A one litre water enema is given two hours before examination.

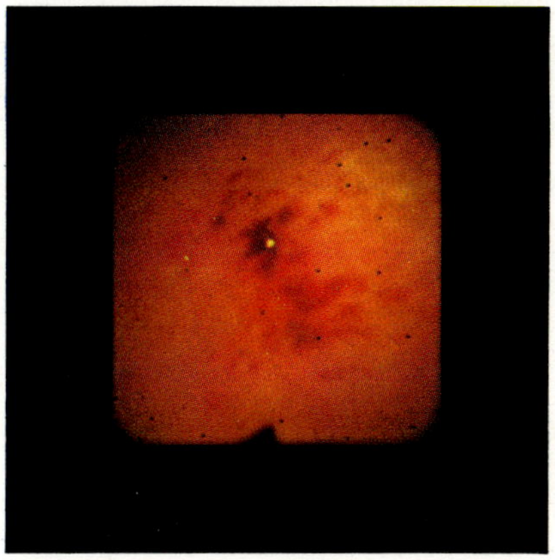

Fig. 2.
Traumatic injury of the mucosa due to prolonged aspiration of faecal remnants. The smooth, shiny, greyish-pink mucosa is suffused with blood.

Fig. 3.
Laxative-induced bauhinitis. Endoscopic view through the ileocecal papilla of the mucosa of the terminal ileum. The mucosa is scarlet red, oedematous, congested and apparently infiltrated. Biopsy of the congested area (Fig. 4) only revealed signs of inflammation.

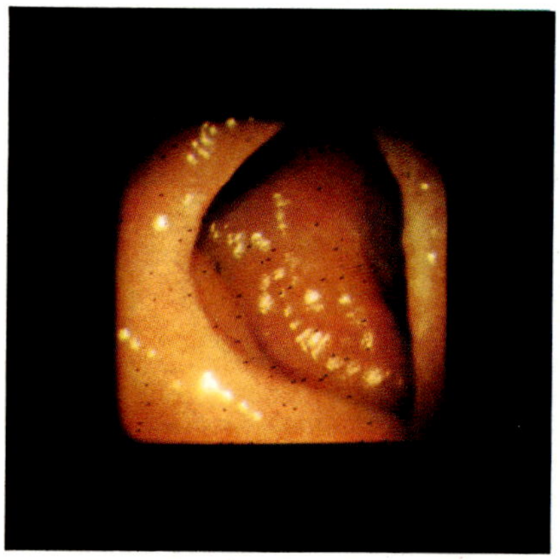

Fig. 3.

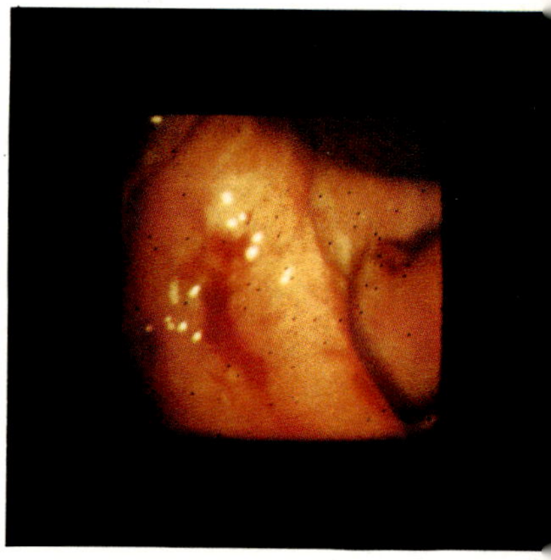

Fig. 4.

'reparation of outpatients for coloscopy:

– 1st day: the patient is allowed 150 ml of milk or tea with one teaspoonful of sugar for breakfast. Lunch and supper together consist of not more than 200 g of grilled meat or fish and two portions of milk pudding. Unlimited amounts of water are allowed. One tablet of Senakot is given every 8 hours.
– 2nd day: the diet is the same as on the lst day. One tablet of Senakot is given every 8 hours. A proctoclysis of one litre of water with a teaspoonful of granular Senakot is given in the morning and in the evening.
– 3rd day: 200 g of grilled meat, one portion of milk pudding, and unlimited amounts of water are allowed. A one litre enema of water with a teaspoonful of granular Senakot is given in the evening.
– 4th day: no food is given but unlimited amounts of water are allowed. A one litre water enema is given two hours before examination.

paghetti, rice, soups of any kind, raw or cooked fruit or 'egetables, wine, spirits, aperitives and coffee are forbidden luring preparation.

Vith these methods, complete coloscopy in the absence of aeces is possible in 95% of cases.

'reparations containing bismuth salts, iron or codeine must ıot be given. Women should not be prepared and examined luring their monthly period but 10 to 12 days before the ıenses.

'he type of preparation can naturally be varied in relation ɔ the clinical and radiological findings. Complete preparation ; unnecessary when cancer of the second portion of the igmoid colon is suspected. In this case a proctoclysis ıorning and evening on the previous day and evacuation of he bowel by means of an enema containing monosodium hosphate, disodium phosphate and sodium benzoate one or wo hours before examination will suffice. The same holds lso in the presence of stricturing of the sigmoid colon or istal portion of the descending colon. In patients with liopathic ulcerative colitis, especially in the active periods

of the disease, preparation must consist only in a 500 ml procotoclysis of normal saline solution at room temperature given a few hours before examination. During remissions of the inflammatory process or when it has become chronic adequate preparation must be carried out according to the bowel function. Patients with diverticulosis of the colon represent a more challenging and complex problem. In these cases the method of preparation adopted in hospitalised subjects should be applied, but the doses of aperients must be considerably reduced to avoid causing diverticulitis.
Adequate cleansing of the colon is very difficult in the presence of a tight high stricture, in the descending or transverse colon, and even meticulous preparation may not be sufficient to clear the large intestine of all the faeces. This should discourage neither the patient nor the endoscopist but should goad them to prolong preparation.
In a patient with a doubtful radiological picture suggesting a tumour of the cecum nine days were required to cleanse the colon and to show up the radiologically doubtful growth.
Preparation need only consist in one day on a fluid diet and two enemas on the night and morning before examination in patients previously subjected to resection of the sigmoid colon in whom the lack of the normal curvature of the sigmoid colon favours evacuation and rapid inspection.
Unexpelled small masses of faecal matter can be removed with the biopsy forceps or by washing out with water injected through the coloscope or through a supplementary teflon tube introduced through the biopsy forceps channel, thus showing up possible underlying lesions. An apparatus for the irrigation of the colon under controlled pressure (Henderson) has recently been introduced with which the colon can be cleansed completely in 4 or 5 hours. This interesting and valuable novelty has certain limitations, like any other method.

Premedication

No medication is usually necessary before coloscopy. No sedatives, analgesics or antispastics have been used in our Department for about three years. In our opinion, adequate psychological management by the nursing staff and, above all, by the endoscopist before and during examination is more useful. The patient should be informed that the examination will not cause pain, but only some distress due to the presence of the instrument in the colon and to the insufflated air. It is also advisable to talk to the patient during examination to reassure him that all is proceeding well.

Sedatives or antispastics need be given only if indispensable or if expressly requested by the patient or when the patient seems unduly restless. Diazepam, 10-20 mg given intravenously or intramuscularly, pethidine, 25-30 mg, or hyoscine-n-butyl bromide, 30 mg, can be used in these cases. Some endoscopists have obtained satisfactory analgesia with the inhalation of nitrous oxide and oxygen (Salmon et al.). General anaesthesia is required in children under 13, whereas only 2‰ of our adult cases were examined under general anaesthesia.

I am personally unfavourable to coloscopy under general anaesthesia because it is essential that the patient should collaborate with the endoscopist by informing him of the onset of pain, a possible warning sign of impending perforation or of a false passage of the coloscope into a diverticular fold.

To conclude, premedication for coloscopy can be schematically defined as follows:

1) no medication — psychological preparation
2) antispastics (Buscopan)
3) sedatives (Diazepam)
4) analgesics
5) general anaesthesia is usede routinely in patients under the age of 15.

Duties of the nursing staff

The nursing staff must

1) arrange admittance of the patient to be examined;
2) prepare the equipment (check the suction apparatus, lubricate the biopsy forceps with liquid paraffin, coat the lenses with Merkril solution, prepare slides for cytology and bottles containing formalin or 70° alcohol for biopsy specimens, check the still and motion picture cameras);
3) place all the equipment required for endoscopic polypectomy on a trolley;
4) place all the patient's records, coloscopy form and X-rays on the endoscopist's desk;
5) explain the technique of coloscopy to the patient and try to place him in a comfortable left lateral position;
6) prepare the emergency equipment (sphygmomanometer, stethoscope, oxygen, first aid drugs);
7) give the sedatives prescribed by the endoscopist;
8) collaborate with the examiner in checking the patient's conditions during examination by taking his pulse rate, by enquiring of his conditions (occurrence of abdominal pain, nausea, etc.) and by reassuring him if restless;
9) take the patient into another room after examination is over;
10) clean the instruments and carefully disinfect the coloscope, biopsy forceps, suction and air channels, etc.

It is important that the endoscopist should always work with the same nurses who will thus become practised in the technique and will come to know the endoscopist habits and fads.

The instrument is usually inserted or withdrawn by an assistant nurse under the examiner's instructions. An intelligent and well-trained nurse will soon become skillful in feeling even the slightest impediment to the progression of the coloscope and immediately inform the examiner.

The nurse will also help the patient to change position as requested by the examiner without displacing the endoscope.

A perfect result depends on adequate collaboration between the patient, nursing staff and endoscopist.

Types of coloscopes

Fig. 5.
Four-way angulation of the bending section of the coloscope.

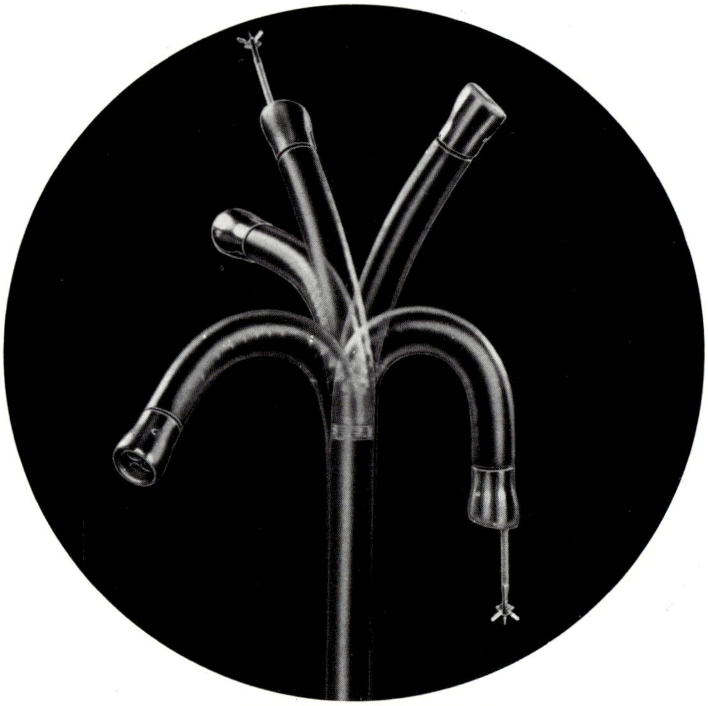

Fig. 6.
Bending section and distal end.

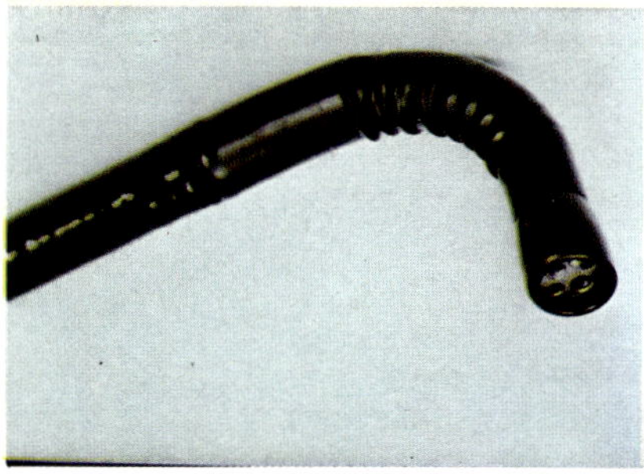

Fig. 7.
Measuring device
Biopsy forceps
Cytology brush.

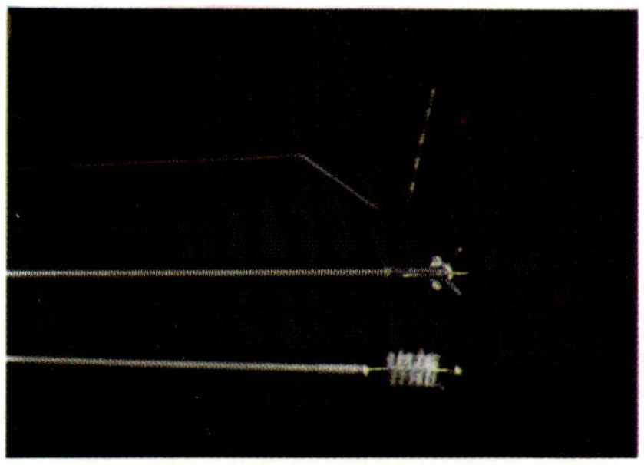

Fig. 8.
Spiked biopsy forceps.

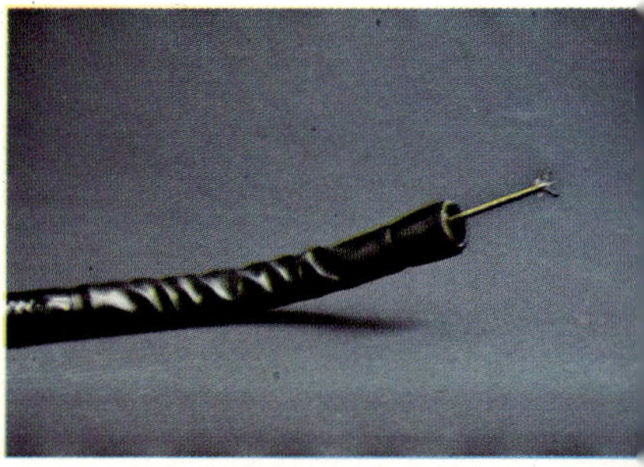

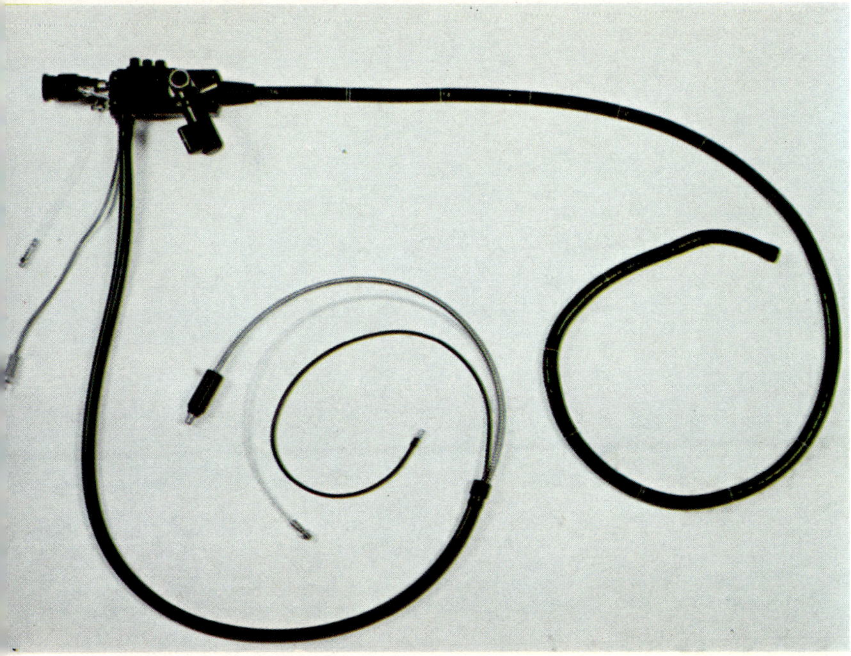

Fig. 9.
ACMI F9-A Operating coloscope.

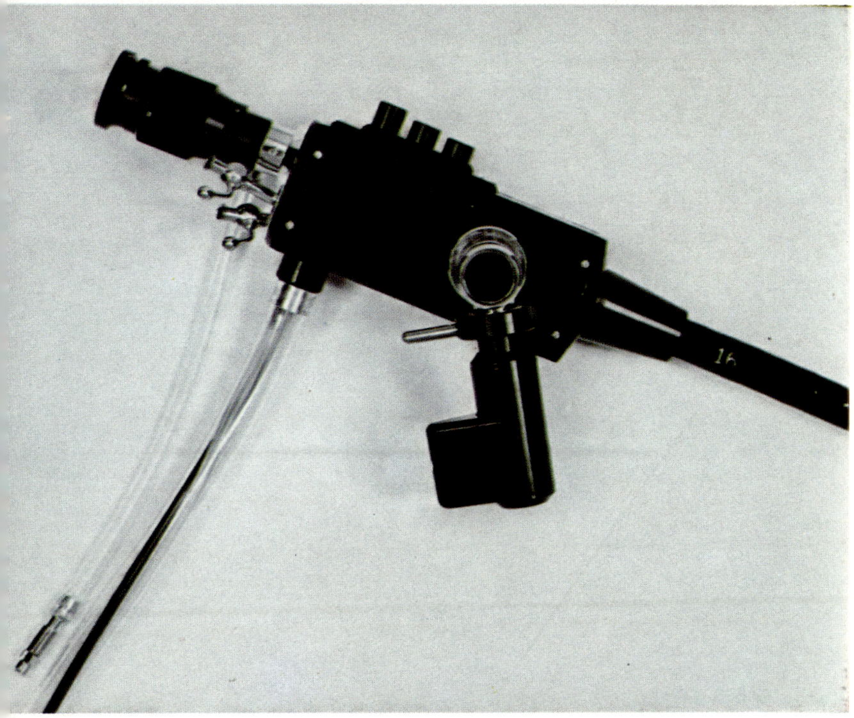

Fig. 10.
ACMI F9-A Operating coloscope.
Detail of the proximal extremity.

ig. 11.
ytology brush.

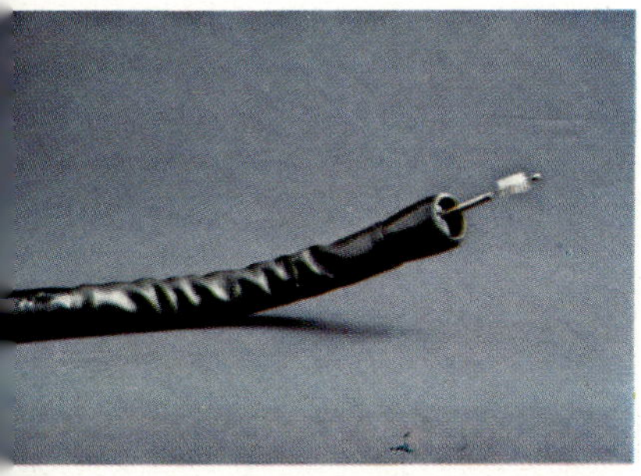

Fig. 12.
Measuring device.

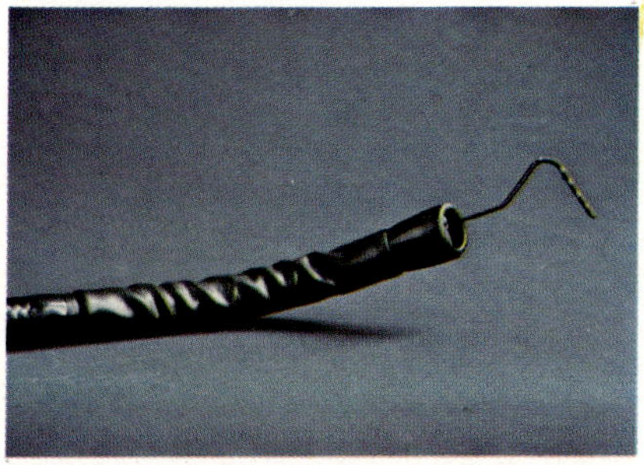

Fig. 13.
Teaching attachment.

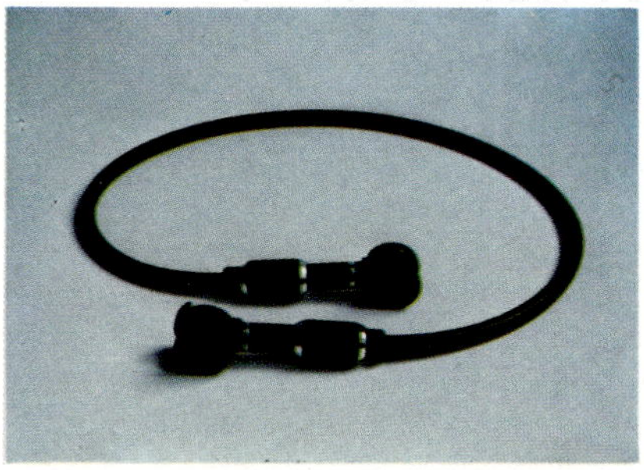

Fig. 14.
Cold Pight source
and accessories.

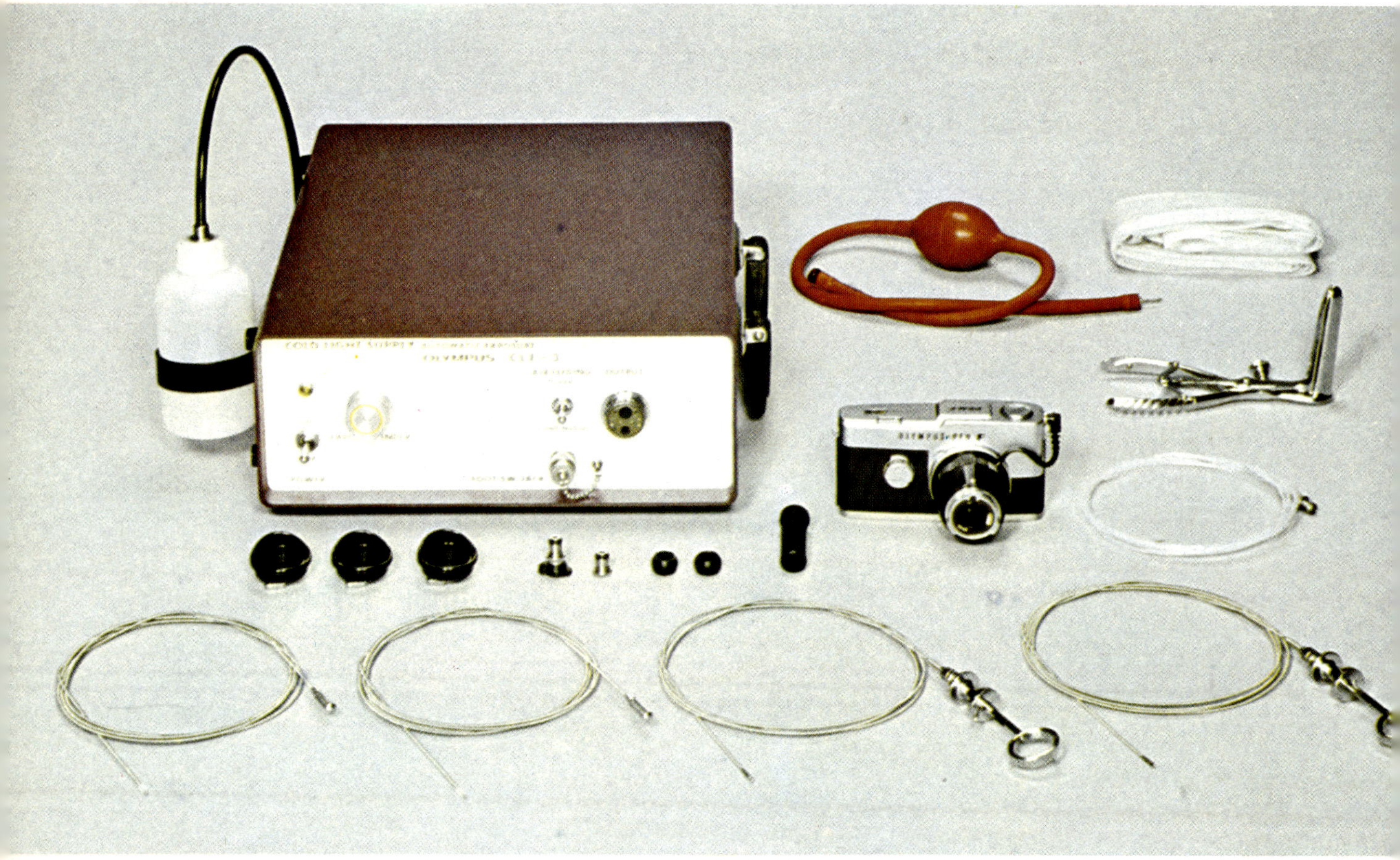

Fig. 15.
Biopsy forceps.

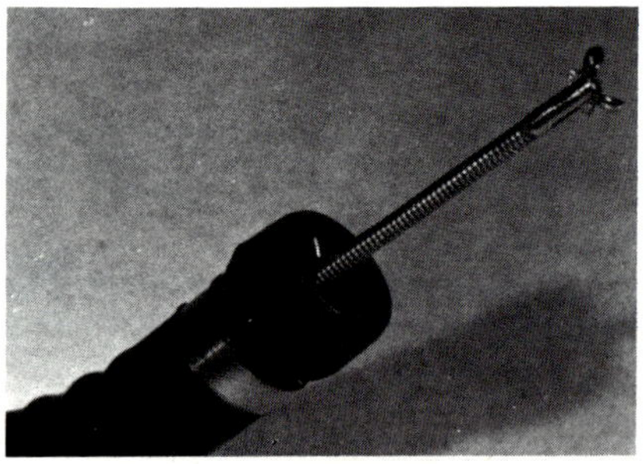

Fig. 16.
Polyethylene water jet tube.

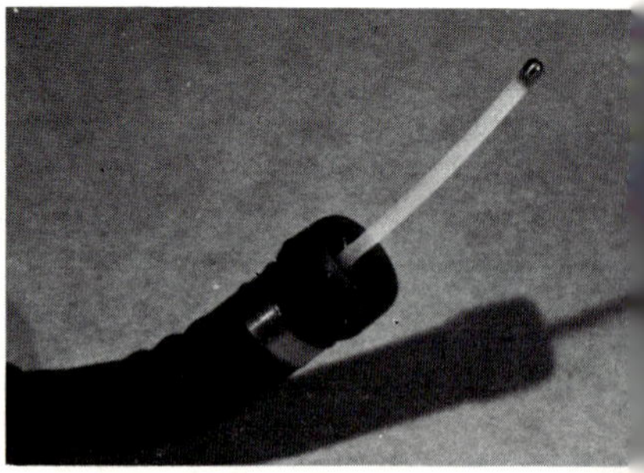

ig. 17.
)lympus Colofibrescopes
Model CF types MB and LB.

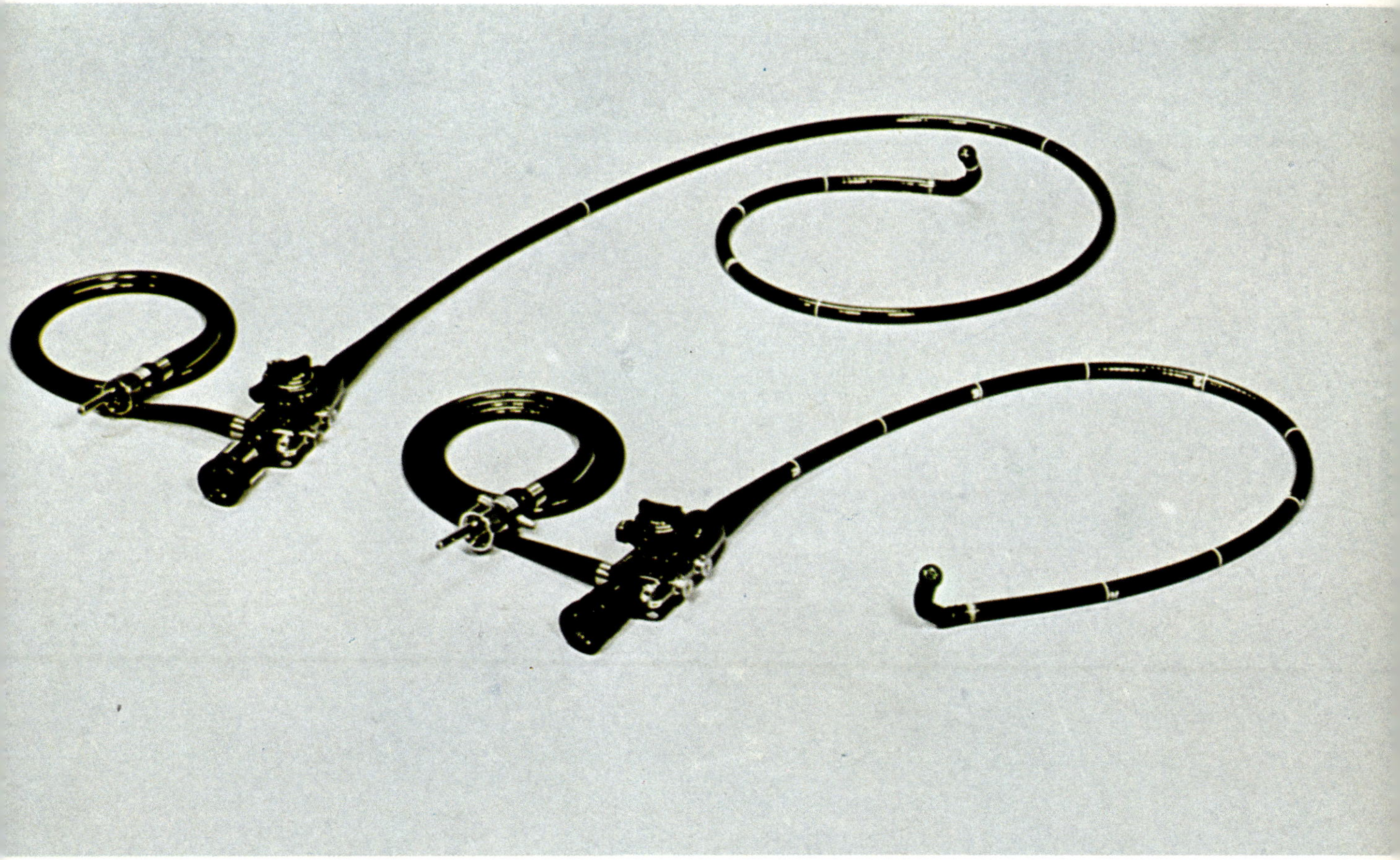

ig. 18.
ytology brush.

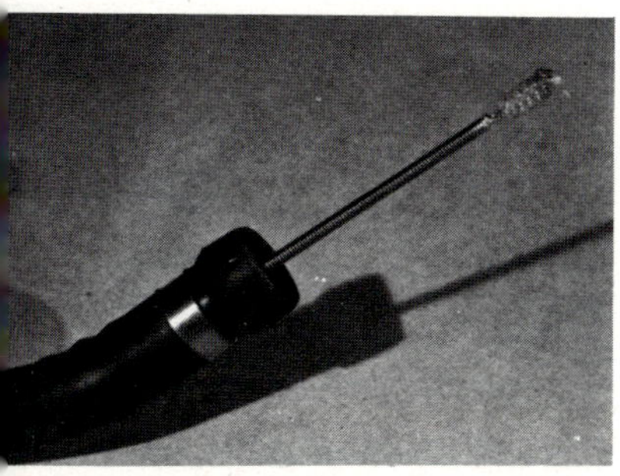

Disinfection of the fibrecoloscope

isinfection of the fibreoptic coloscope, recently still reputed minor question, has yet to be perfected and, on the contrary, epresents an important problem. A disinfectant capable of estroying the varied and resistant intestinal flora on the oloscope without damaging this delicate instrument has so ar not been found.

Ionetheless, it is indispensable that the coloscope be isinfected to free it as thoroughly as possible of the bacteria linging to its sheath and, especially, tip and inside the suction/ iopsy and water channels. A chloroacetic acid derivative r glutaraldehyde have been used with satisfactory results by ome endoscopists whereas others favour the use of a gas such s ethylene oxide.

'he best method is perhaps to first wash the fibrescope with vater and disinfectant solutions and then place it in an irtight plastic or metal sheath into which the gas is introduced t room pressure.

'airly satisfactory disinfection, although not perfect terilization, can be obtained with this method.

ACMI have recently introduced a pratical and effective utomatically programmed steriliser comprising the GS-10 terilizer and the GA-11 Aerator. The endoscopes and ncillary equipment are placed in the GS-10 chamber where hey are treated with ethylene oxide from a disposable artridge. The machine can be programmed for one of two ycles: sterilisation lasting 64 minutes, or disinfection lasting 6 minutes. The instruments are then transferred for 2 hours nto the GA-11 Aerator to remove the ethylene oxide residues.

A sterilising activity of ethylene oxide has been documented gainst Staphylococcus aureus, the Salmonellae, Pseudomonas eruginosa, Mycobacterium tuberculosis and others, not gaints hepatitis virus.

No damage to the endoscopes is caused by ethylene oxide at noderate temperatures and at normal pressure.

To summarize:

1) the outer sheath is first washed with neutral soap and warm water;
2) the outer sheath and channels of the instrument are then washed for at least 5 minutes with an antiseptic solution (Cidex, Hisohex, Alhidex) using special connections;
3) the outer sheath is again washed with 20% alcohol;
4) the instrument is then sterilised by means of a gas (ethylene oxide) or mixture of gases.

Method of examination

Fig. 1

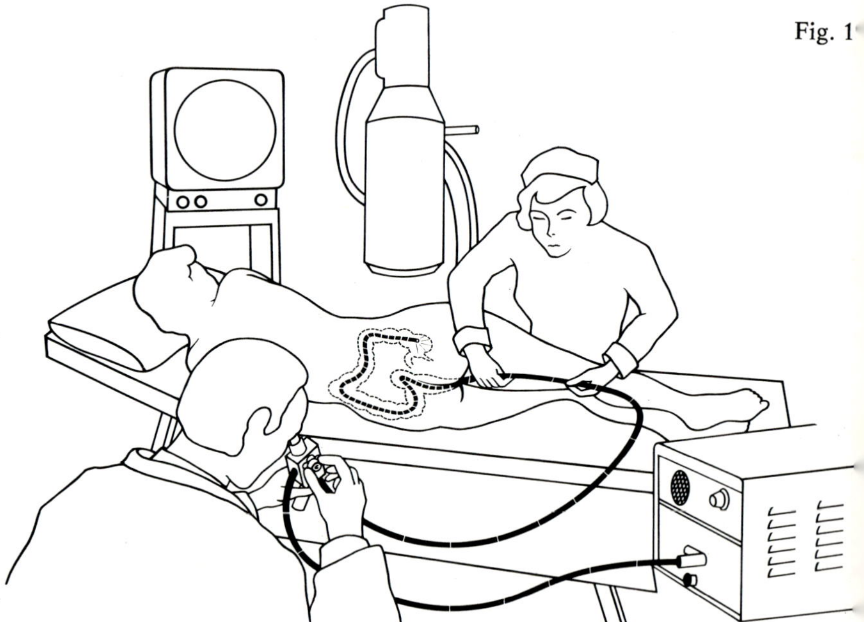

he endoscopist should first carefully check all the equipment ıd examine the patient's records and the results of ıdiological examination by barium enema to become ;quainted with the diagnostic problem. The most significant -rays are placed on the X-ray viewer. Radiological ‹amination of the colon by barium enema prior to)loscopy is absolutely essential.

he patient is then examined for surgical scars, laparocele, ıguinal, femoral or umbilical hernia, meteorism, ascites and ɔdominal masses.

he patient is placed in the left lateral position (Figs. 19 & 20). ectal examination is performed and the anal orifice is ıbricated with 1% xylocaine.

he tip of the coloscope, also lubricated with xylocaine, is ıen inserted through the anus and air is insufflated to distend ıe rectum. The coloscope is then directed towards the :ctosigmoid junction with the lumen under constant visual ɔntrol. The coloscope can be steered through the loops of ıe sigmoid colon by working the controls on the near end of ıe instrument to manoeuvre the tip, and by rotating the shaft .ockwise or counterclockwise as required. Blind insertion ıould be attempted only for short distances and provided ıe mucosa can be seen to move along under the tip of the ɔloscope. The coloscope must be withdrawn at once to avoid ıe risk of perforation if the mucosa whitens and thins and ıe insufflated air is seen to rebound from the distended owel wall, or when the patient complains of sudden pain.

g. 20.

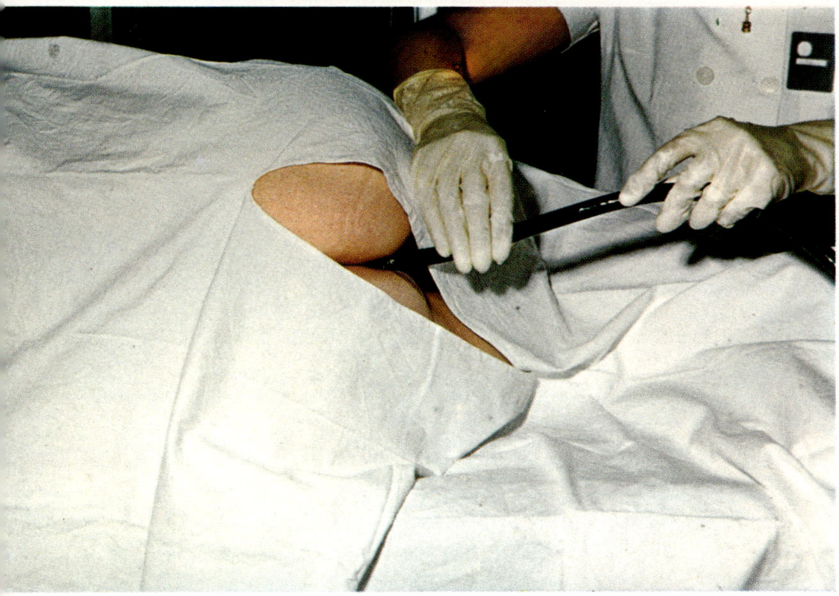

Figs. 19 & 20.
The patient is in the left lateral position with the examiner standing behind him. The nurse faces the patient holding the coloscope in her left hand while using her right hand to insert the instrument according to the examiner's instructions.

If the coloscope fails to pass the first bends of the sigmoid colon, rather than renewing attempts to insert the instrument, the patient should be placed in a supine position and the exact position of the coloscope should be checked fluoroscopically (Fig. 21).

Abdominal palpation may be helpful although it is usually ineffectual in negotiating the sigmoid colon. When insertion, rotation, abdominal palpation and change of position have all been attempted to no avail, the tip of the coloscope can be hooked to a fold after having aspirated enough air to cause the sigmoid colon to collapse. The shaft of the coloscope is then rotated counterclockwise and simultaneously withdrawn thus pulling the sigmoid colon downwards and straightening out the bends which impeded progression of the instrument towards the descending colon. The coloscope must not be pushed forwards when it is looped up in the sigmoid colon because progression would become more and more difficult and painful and would eventually cause impaction of the tip and make it impossible to manoeuvre it by means of the external controls. Loops must be straightened out as they form.

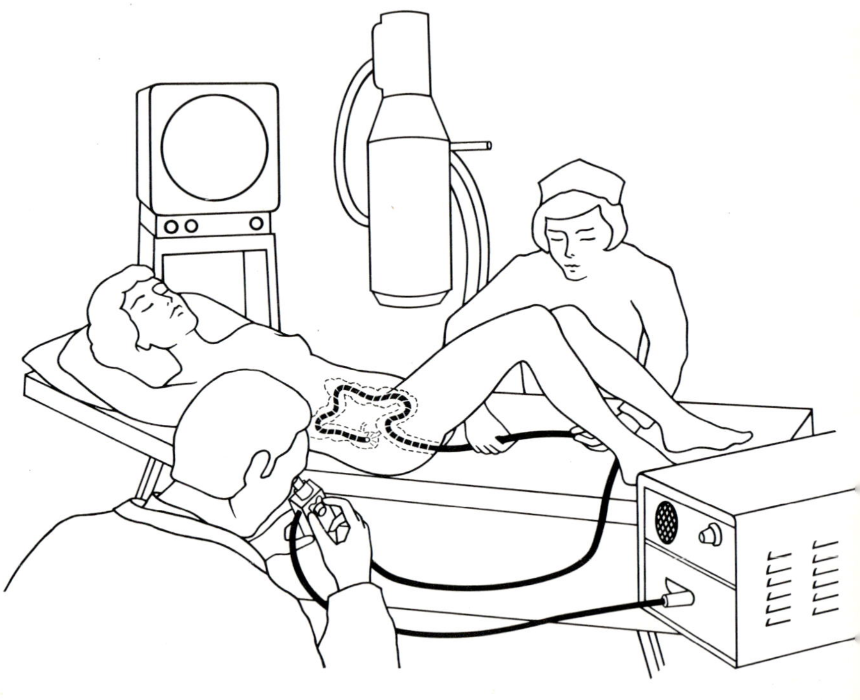

Fig. 21.
The patient has passed from the left lateral to the supine position to enable the progression of the coloscope to be checked fluoroscopically with image intensification. The examiner and nurse stand as before.

'igs. 22, 23, 24 & 25 show how the coloscope can easily ass through the sigmoid colon into the descending colon. 'he main difficulties encountered in coloscopy are in egotiating the sigmoid colon and, especially, the sigmoid-escending colon junction.
f the tip of the coloscope is now pointed in the correct irection towards the descending colon it will automatically dvance towards the splenic flexure provided the sigmoid olon is straightened out by pulling on the shaft of the oloscope and rotating it clockwise. When positioned as in 'ig. 28 the tip of the coloscope may become impacted nder a haustral fold instead of moving on towards he descending colon. In this case the sigmoid colon is isplaced upwards and the tip is arrested between the austrations. This is indicated by difficulty in advancing he coloscope and can be confirmed by fluoroscopy. The oloscope must then be withdrawn until the sigmoid colon is traightened out and its tip must then be aimed at the escending colon.

Fig. 22.

Fig. 23.

Fig. 24.

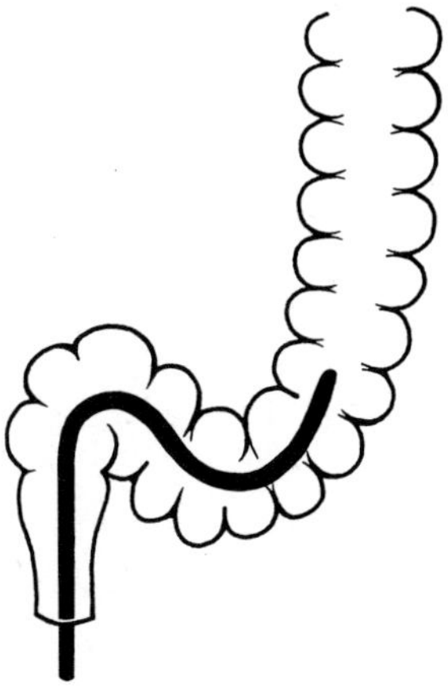

Fig. 25.

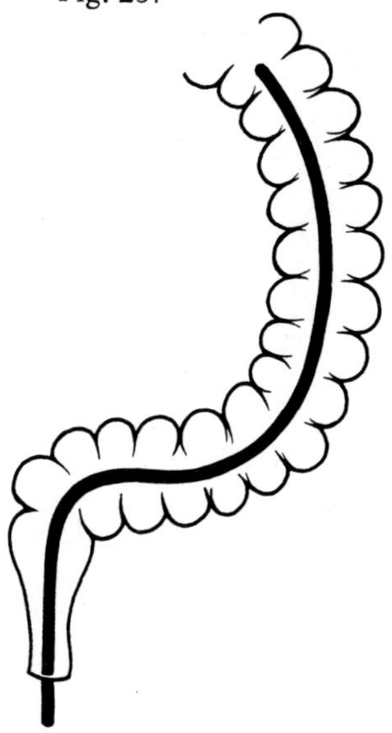

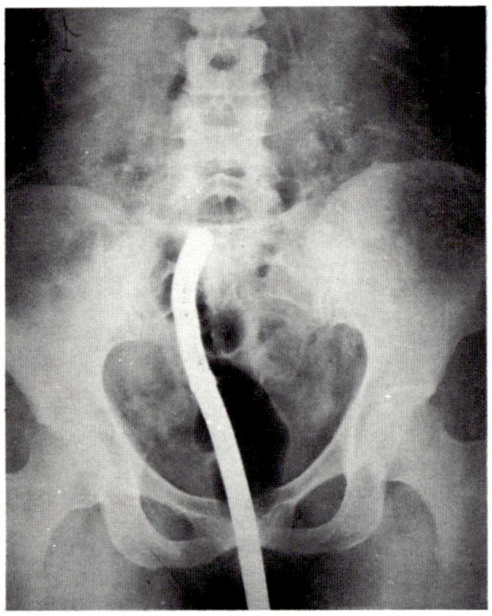

Fig. 26.

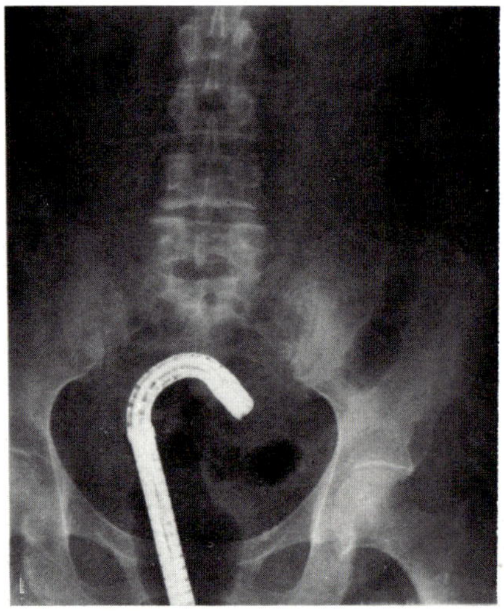

Fig. 27.

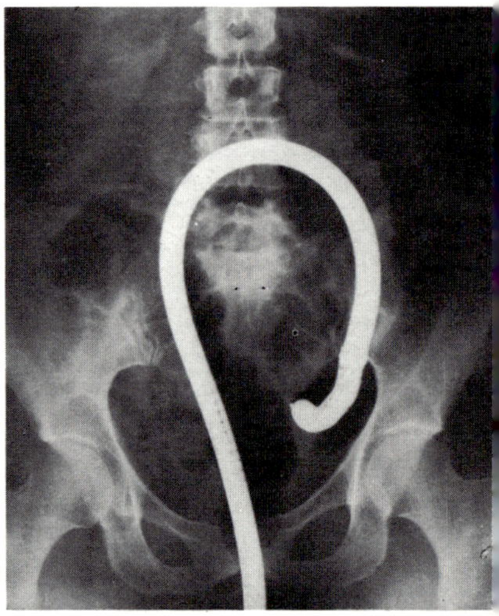

Fig. 28.

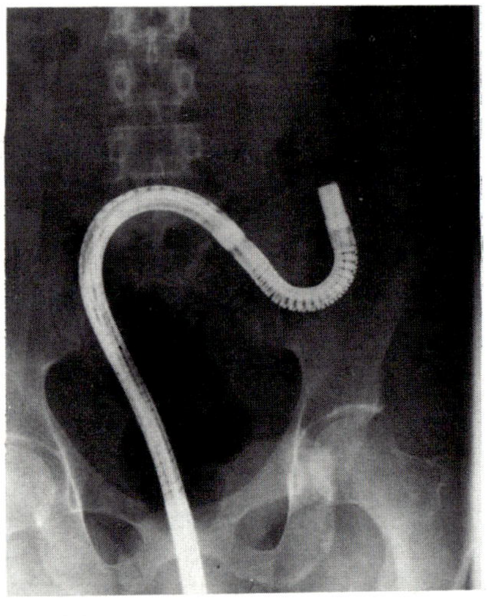

Fig. 29.

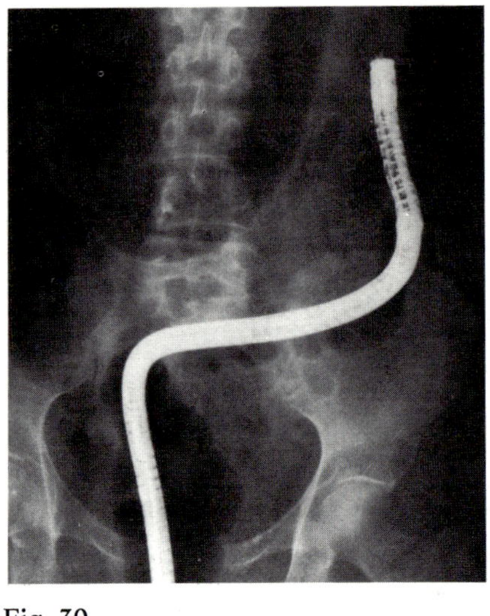

Fig. 30.

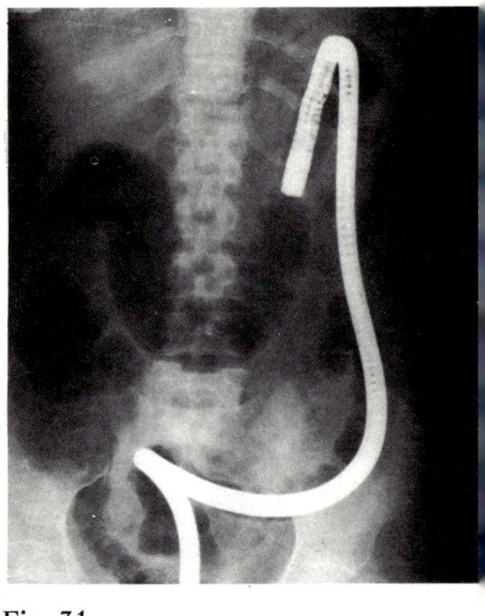

Fig. 31.

Progression of the coloscope in the sigmoid and descending colon

The flexible coloscope passes through the rectosigmoid junction (Fig. 26) into the middle portion of the sigmoid colon (Fig. 27) to reach the sigmoid-descending colon junction where the tip becomes impacted (Fig. 28). The tip of the coloscope is then flexed to face the descending colon and the instrument is withdrawn and slightly rotated clockwise until it gains the descending colon (Fig. 29). The sigmoid colon is straightened out (Fig. 30) and the tip of the coloscope can then be easily pushed through the left flexure of the colon (fig. 31).

The tip of the coloscope may become impacted at the sigmoid-descending colon junction and all attempts to steer in into the descending colon will then be of no avail (Fig. 32). In this case the coloscope should be withdrawn into the first portion of the sigmoid colon, just past the rectosigmoid junction (Fig. 33), and then rapidly and forcibly rotated counterclockwise (Fig. 34).

This displaces the sigmoid colon from left to right (Fig. 36) and automatically exposes the lumen of the descending portion of the sigmoid colon. The coloscope is then pushed into the descending colon forming an alpha loop (Figs. 37 & 38). The coloscope will then advance without further hindrance as far as the splenic flexure. Once the acute pain caused by the drastic rotation of the coloscope has subsided the patient feels no discomfort except for slight suprapubic or epigastric distension according to the width of the alpha loop.

Once the left flexure is reached the tip of the coloscope is hooked to the first portion of the transverse colon and the instrument is withdrawn and rotated clockwise (Fig. 39) so as to straighten out the loop and shorten the descending colon with the sigmoid colon in a vertical position (Fig. 40). These manipulations must be done under fluoroscopic observation.

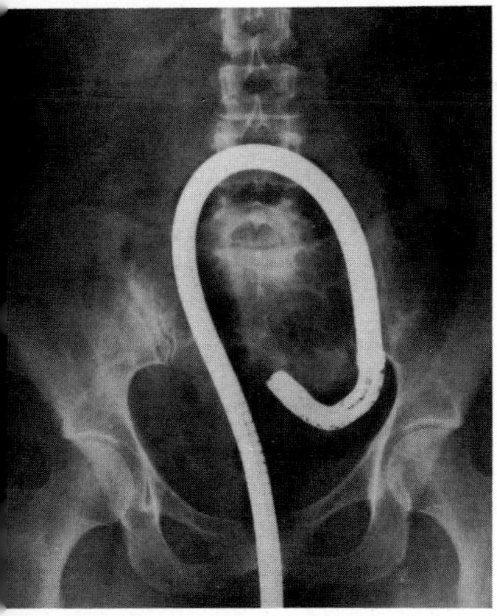

Fig. 32.
The tip of the coloscope is impacted at the sigmoid-descending colon junction.

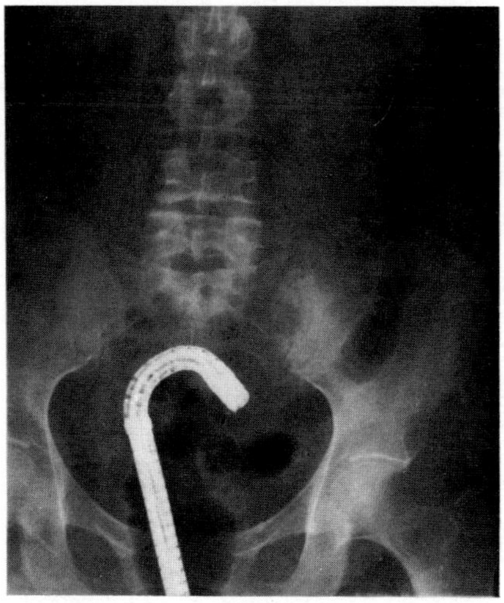

Fig. 33.
The coloscope is withdrawn into the first portion of the sigmoid colon to make rotation possible.

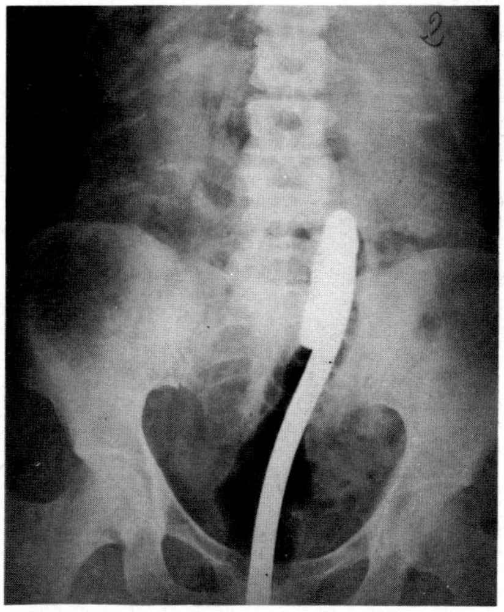

Fig. 34.
The coloscope has been rotated counterclockwise. The sigmoid colon is displaced to the right.

Fig. 35.

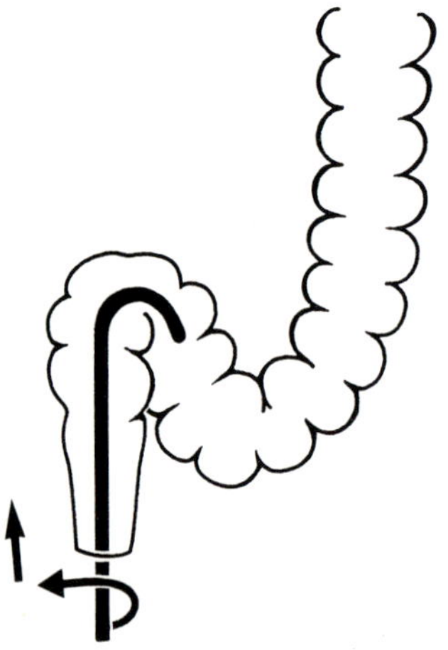

Fig. 36.

Fig. 37.

Fig. 38.

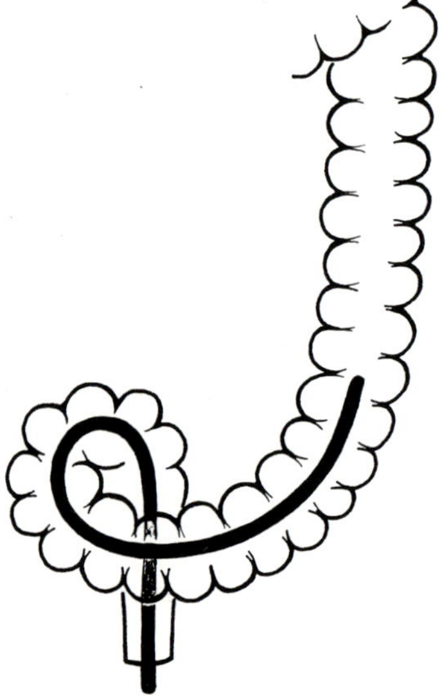

Fig. 39.

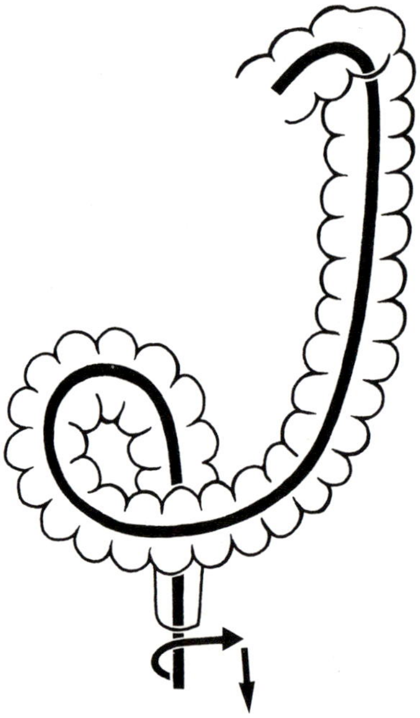

Fig. 40.

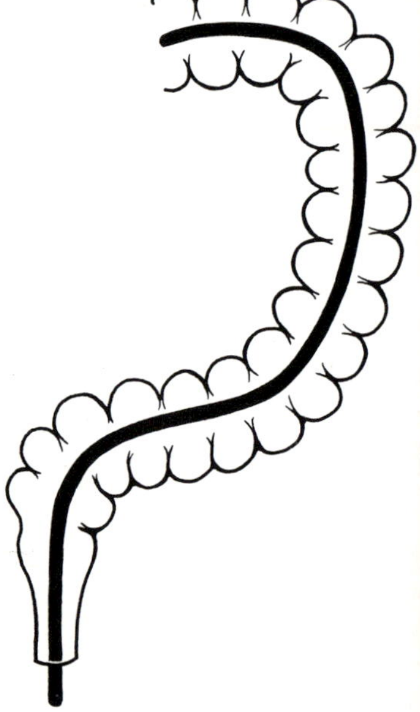

Progression of the coloscope along the large bowel

The fibrecoloscope forms an alpha loop in the sigmoid colon and then descends into the pelvis towards the distal portion of the descending colon (Fig. 41). The loop is partly straightened out by slight clockwise rotation and the coloscope is advanced as far as the transverse colon (Figs. 42, 43 & 44). At this point the loop can be straightened out completely by forcibly pulling the shaft downwards and rotating it clockwise after which the coloscope can be advanced as far as the cecum (Fig. 45). By withdrawing the fibrescope the sigmoid colon can be shortened and straightened out. The instrument and large intestine then take on the shape of a wide question mark (Fig. 46).
Once the whole sigmoid colon has been straightened out and

Fig. 41.

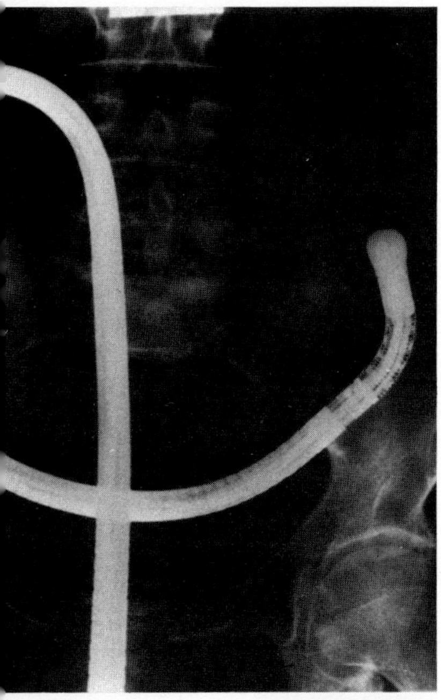

Fig. 42.

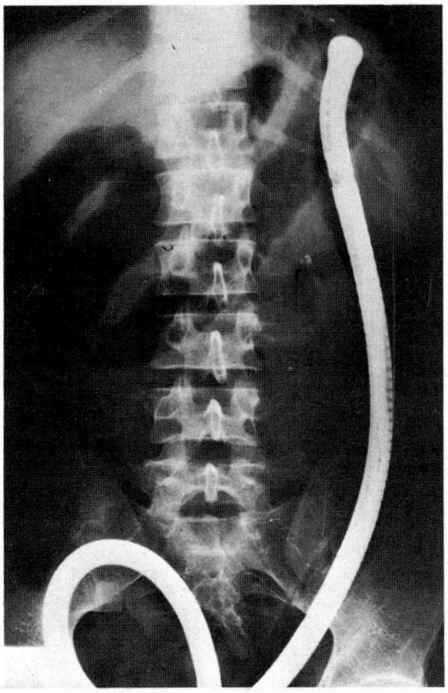

Fig. 43.

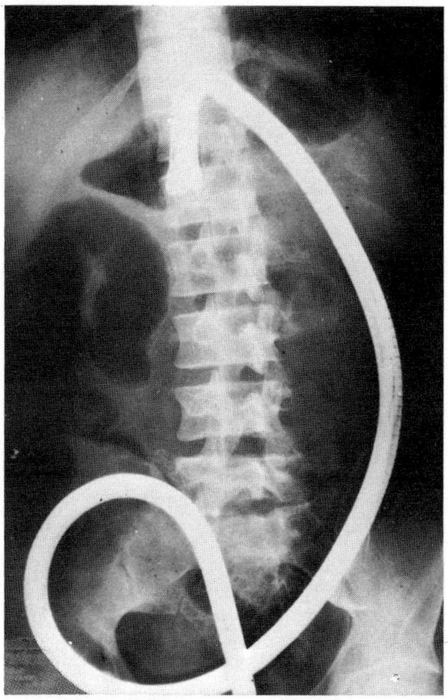

Fig. 44.

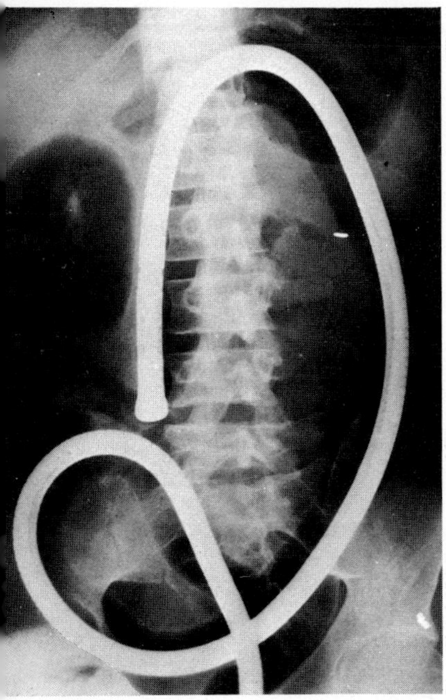

Fig. 45.

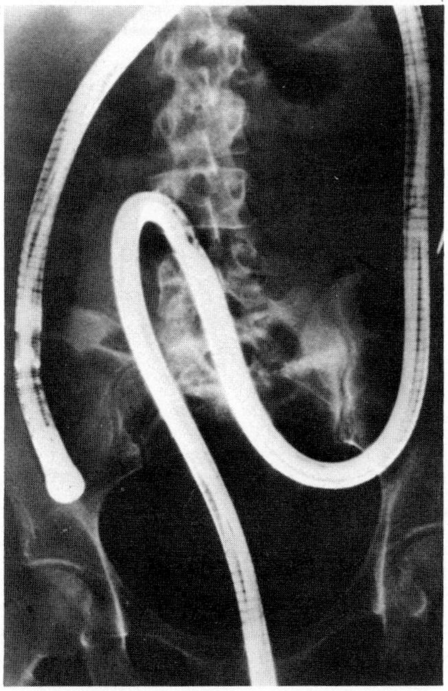

Fig. 46.

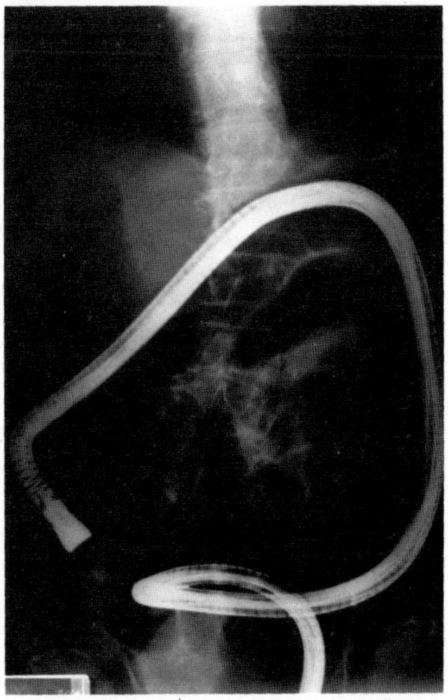

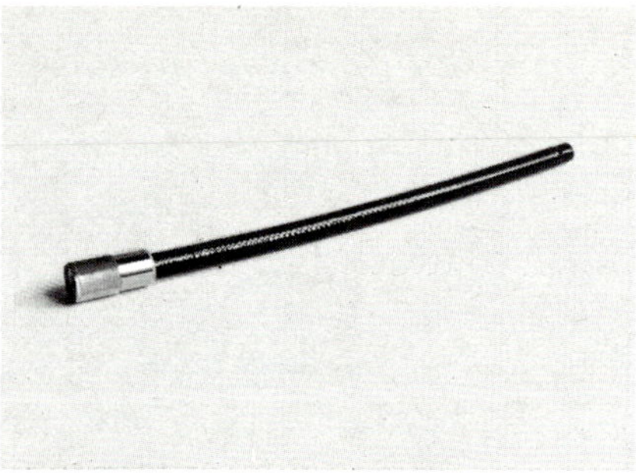

Fig. 47.
Olympus stiffening tube with metal connection.

the descending colon has been shortened, examination can proceed speedily. Other loops often form as the coloscope passes along the transverse colon, as indicated by increasing difficulty in advancing the fibrescope. In this case fluoroscopy will also be very helpful. To avoid this drawback, which increases the time required to complete examination, the fibrescope should be stiffened either by means of an external stiffening tube formed by a plastic-coated wire spring (Fig. 47) or by inserting a steel wire in the biopsy forceps channel. The latter device has the drawback that it makes aspiration impossible (the stiffener completely obstructs the suction channel) and entails the risk of perforation of the plastic lining of the biopsy forceps channel during insertion, but offers the advantages that it can be inserted even when the fibrescope is not perfectly straigth, that it can be used in conjunction with an external stiffening device and that it also stiffens the tip.
The use of a stiffening device is required in 90% of coloscopies.
External stiffeners must be applied around the coloscope before starting examination. Great care must be taken not to stretch or crease the rubber sheath around the tip of the fibrescope.
The stiffening tube should be inserted only after the sigmoid colon has been completely straightened out.
With the aid of a stiffening device the coloscope can be made

Fig. 48.
Angulation of the fibrescope at the rectosigmoid and sigmoid-descending colon junctions. A stiffening tube should not be used in these circumstances because of the risk of lacerating the sigmoid colon.

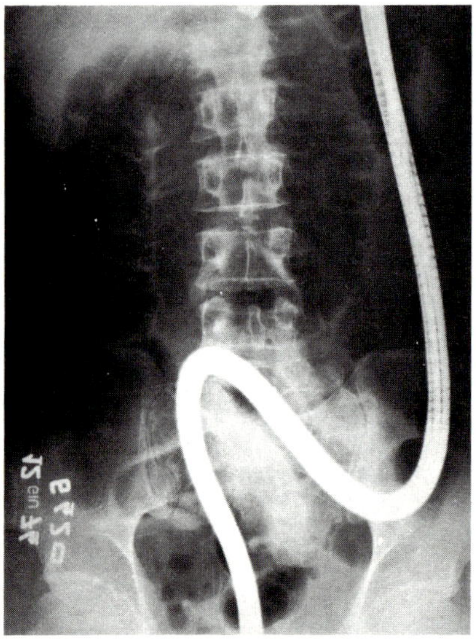

Fig. 49.
The sigmoid colon has been straightened out and forms a continuous channel with the descending colon. A stiffening tube can be inserted without fear.

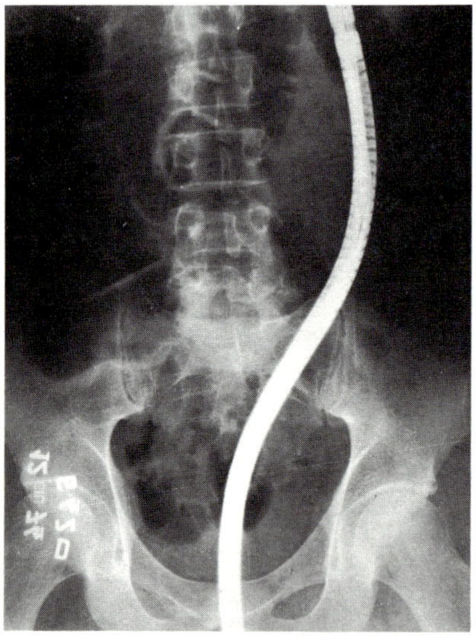

Fig. 50.
The stiffening tube has been inserted and reaches the upper third of the descending colon.

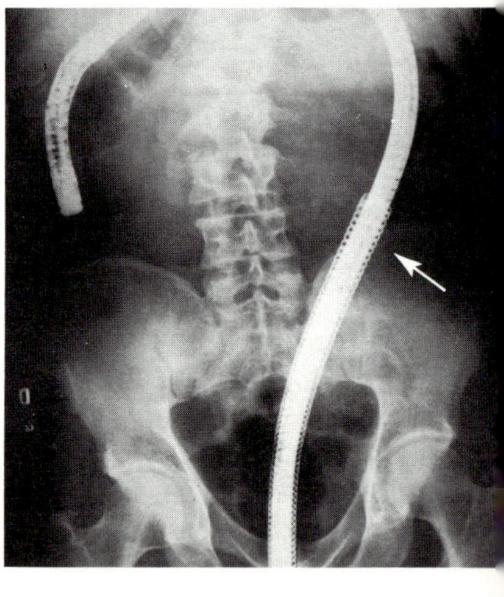

:o advance easily and speedily as far as the middle third of :he transverse colon. Here progression of the coloscope often slows down and the tip may even slide back on account of an ıpward displacement and increased angulation of the ,plenic flexure. In this case the tip of the coloscope should be ıooked upwards and the instrument withdrawn. This nanoeuvre lowers and reduces the angulation of the splenic lexure, lifts and straightens out the transverse colon and visualizes the hepatic flexure.

Once it has reached the hepatic flexure the coloscope may pass into the ascending colon and reach the cecum with no great difficulty. In other instances completion of the examination represents a serious problem which can nonetheless be solved in 95% of cases.

Difficulty in reaching the ascending colon may be due to excessive angulation of the hepatic flexure, redundant loops, excessive mobility of the hepatic flexure, or redundant haustral folds.

Although there is no specific method by which the hepatic lexure can be passed in every case, these difficulties can usually be overcome by changing the patient's position, aided by abdominal palpation, which can sometimes direct

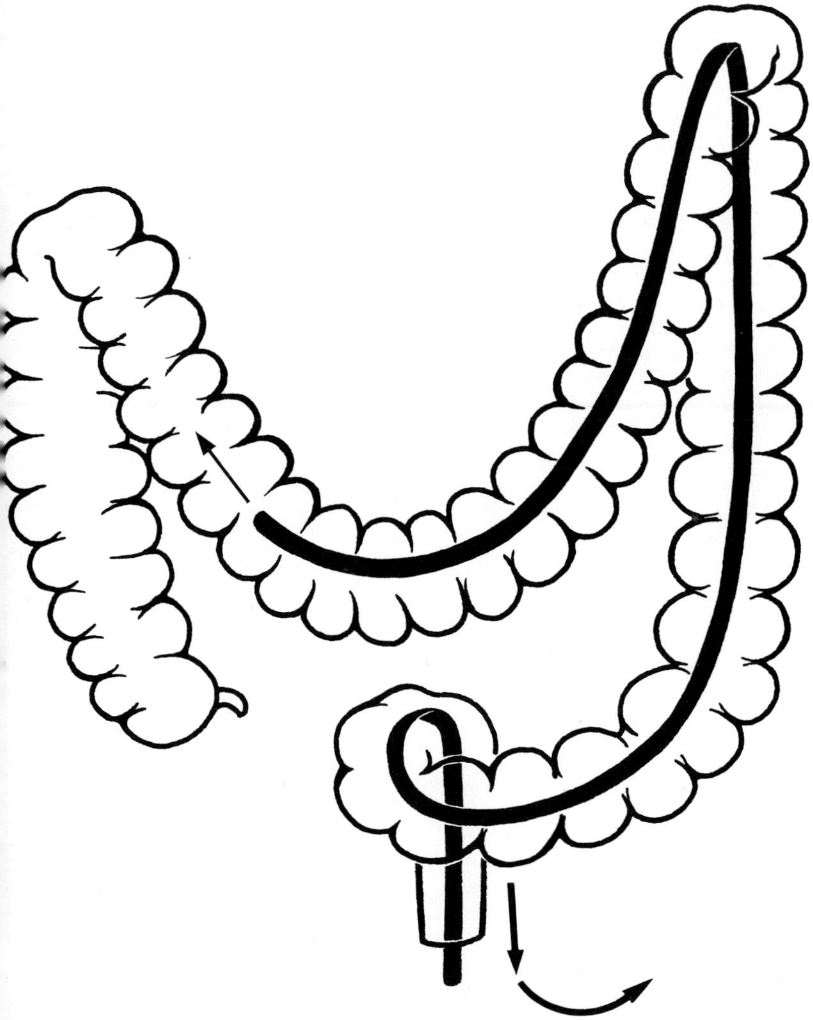

Fig. 51.
Insertion of the coloscope along the transverse colon.

the instrument correctly, and especially by the use of the utmost patience and persistence by the examiner, an indispensable virtue in coloscopy. There are, however, certain conditions such as adhesions and tight strictures which may make coloscopy impossible.

Once it has reached the cecum the coloscope and large bowel form a wide arc (Figs. 56 & 57).

Having thus completed the first part of the examination, taken up almost entirely by technical problems, the examiner can now begin systematic inspection of the mucosa.

The coloscope is withdrawn very slowly, taking care that every single haustration is examined. Rotation to the left or right may be helpful, as well as repeated insertion and withdrawal when the examiner fears that an area has not been examined thoroughly. In most cases every part of the large bowel can ben examined.

Progression of the coloscope along the transverse colon

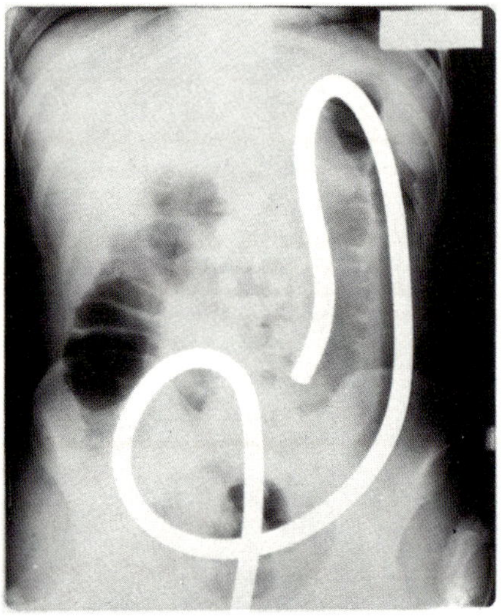

Fig. 52.
The coloscope is in the middle third of the transverse colon. Its progression is slowed down by the upward displacement of the splenic flexure and by a loop in the sigmoid colon.

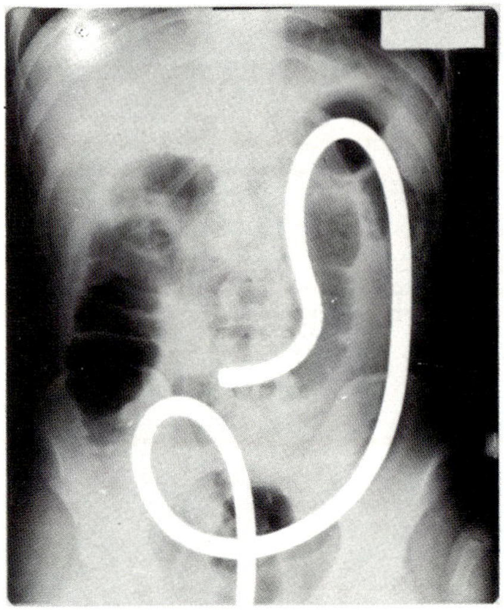

Fig. 53.
The tip of the coloscope is hooked upwards. The instrument can thus be withdrawn and rotated clockwise to straighten out the alpha loop, lower the splenic flexure and raise the transverse colon.

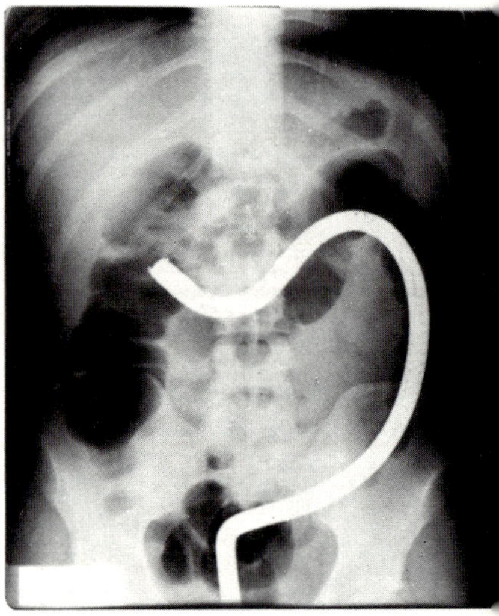

Fig. 54.
The tip has reached the hepatic flexure.

Fig. 56.

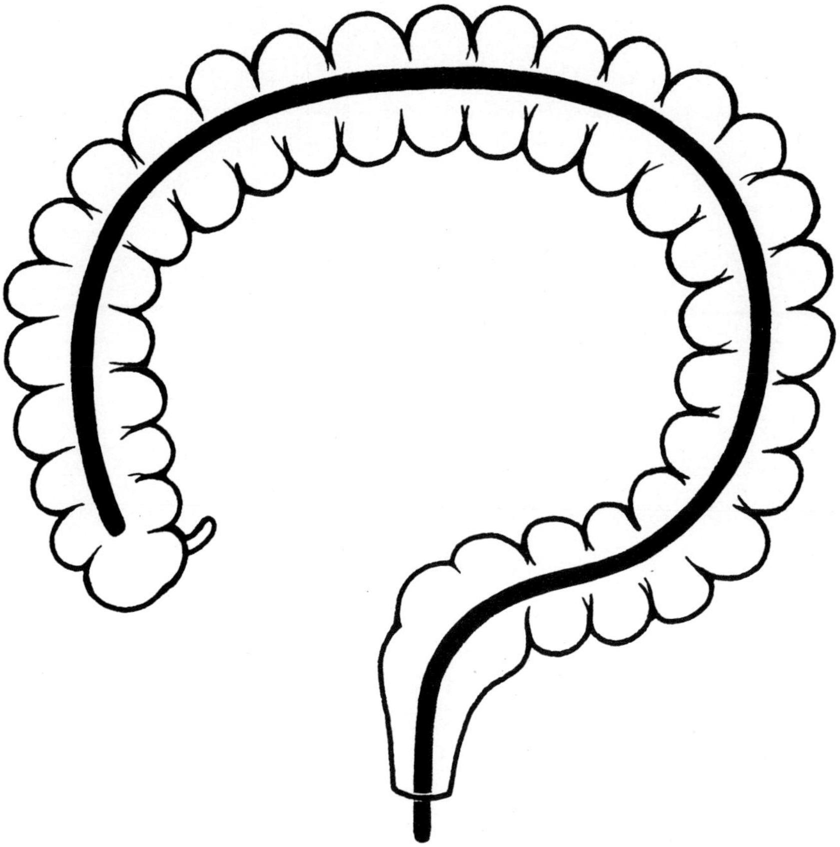

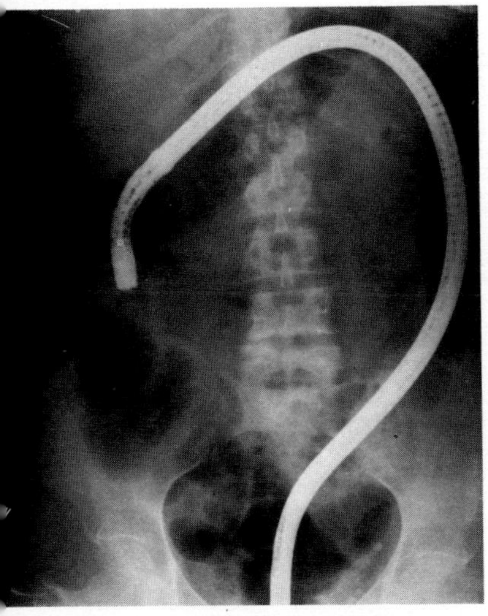

'ig. 55.
The coloscope has passed the
ıepatic flexure and is bent down
acing the ascending colon.

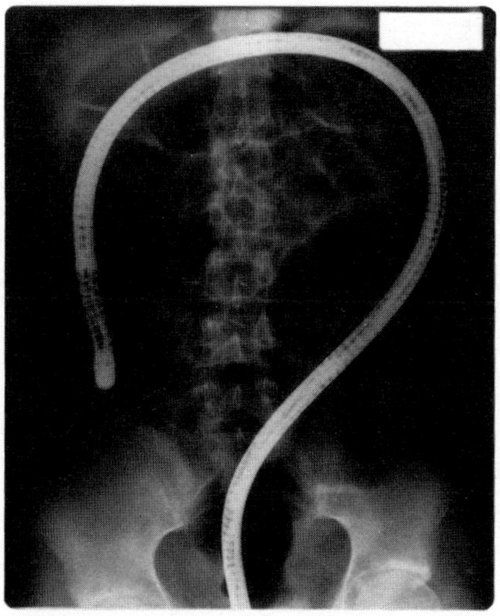

Figs. 56 & 57.
The fibrecoloscope has reached
the cecum.

However, there are certain small areas, at the rectosigmoid junction, sigmoid-descending colon junction, splenic flexure, middle third of the transverse colon, and hepatic flexure, which may escape observation, especially those at the flexures where the coloscope may spring back suddenly on withdrawal. These blind spots on coloscopy can be reduced 70% by using fibrescopes with 120° four-way tip angulation and by suitable manipulation (rotation, repeated insertion and withdrawal) as described above.

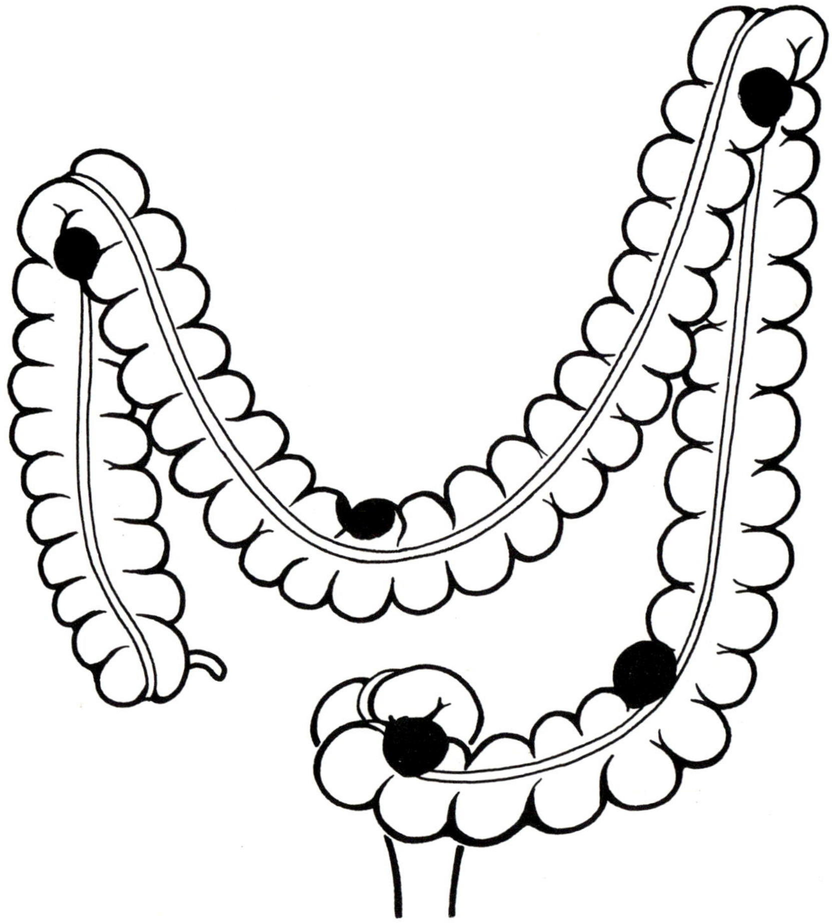

Fig. 58.
Possible blind spots during coloscopy. Their position is variable, according to the morphology of the large bowel.

coloscopic biopsy

\imed biopsy of any part of the large bowel and terminal leum is possible with all types of fibrecoloscope.
The biopsy forceps is inserted and pushed out through the iopsy channel.
t is then directed onto the spot to be examined by pplying to-and-from movements to the forceps itself and by uitable angulation of the fibrescope tip by means of the xternal controls.
There may be difficulty in passing the forceps when the haft of the coloscope forms a tight loop or when the tip is very acutely angulated. Adequate lubrication of the forceps nd a neutral position of the external controls will make it ossible to pass the mobile part of the fibrescope.
The amount of biopsied material is small (1-2 mm) and may ot suffice for complete histologic examination. Biopsy must lways be performed under visual control. Care must be aken not to insert the jaws of the forceps too deeply and ot to perform biopsy if the mucosa appears to be distended y an excess of insufflated air. Perforation may take place f a biopsy is done on a brittle, ulcerated mucosa or on a leeding growth even if the forceps jaws do not pass the nuscularis mucosae (Fig. 61).

Figs. 59 & 60.
Biopsy forceps protruding from the tip of the fibrescope. In Fig. 59 the tip of the fibrescope is in the ascending colon. In Fig. 60 it is at the splenic flexure.

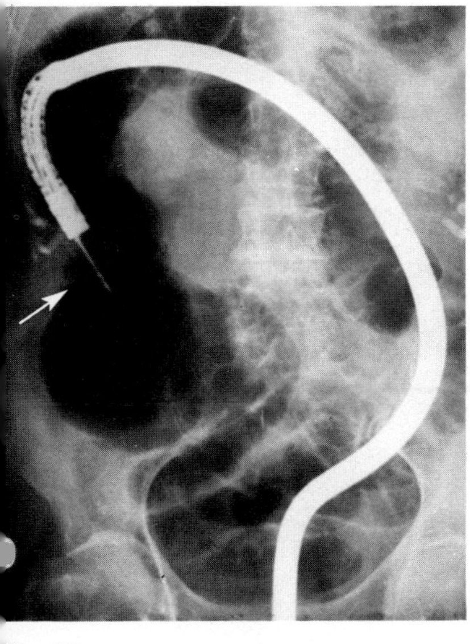

ig. 59.

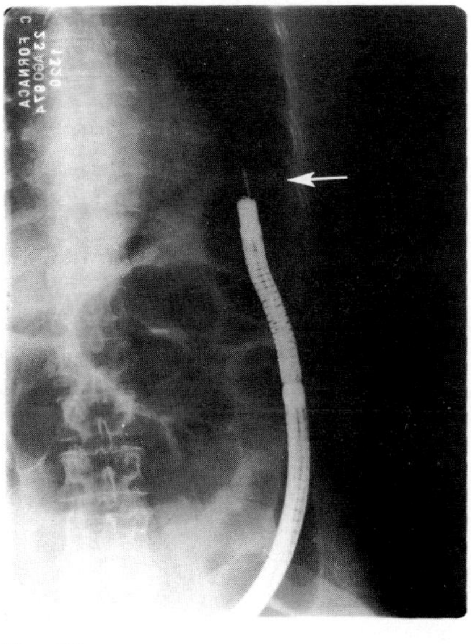

Fig. 60.

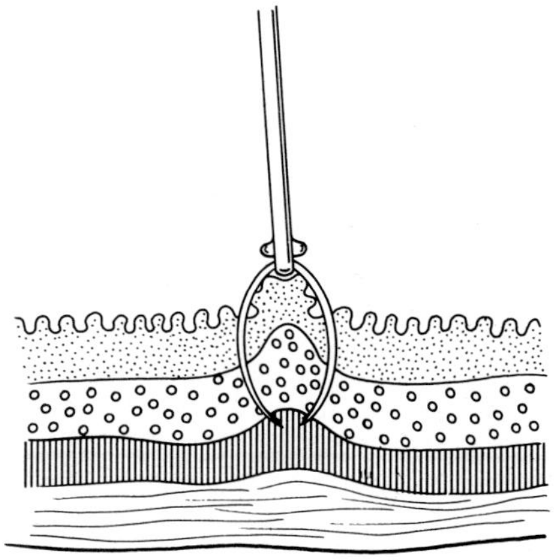

Fig. 61.

Biopsy with the diathermy snare (big particle biopsy)
Larger biopsies can be taken with the diathermy snare. With this method material can be removed from the ubmucosal layer when this is desirable (pseudopolyps, uspected lymphomas, etc.).

coloscopy with abdominal transillumination

In thin subjects the progression of the coloscope can be followed with the aid of transillumination through the abdominal wall. The lighted tip can sometimes indicate the exact position of the coloscope, although this method cannot wholly replace fluoroscopy.
Fluoroscopy should be minimized to avoid damage to the fibre bundle. Bursts of 20 seconds are usually sufficient to determine the exact position of the fibrescope, with a total of about 150 seconds per examination.
By transillumination through the abdominal wall it is possible to determine the position of the tip of the coloscope.
Transillumination is impossible only when tip of the coloscope points towards the patient's back and where the colon is retroperitoneal.

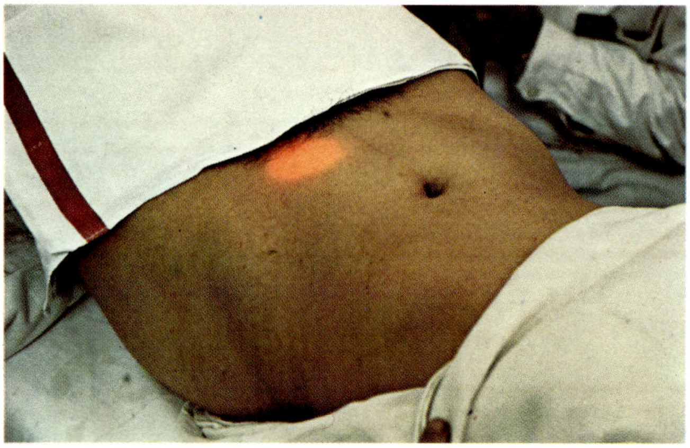

Fig. 62.
The tip of the coloscope is in the left iliac fossa, i.e. in the middle-distal portion of the sigmoid colon.

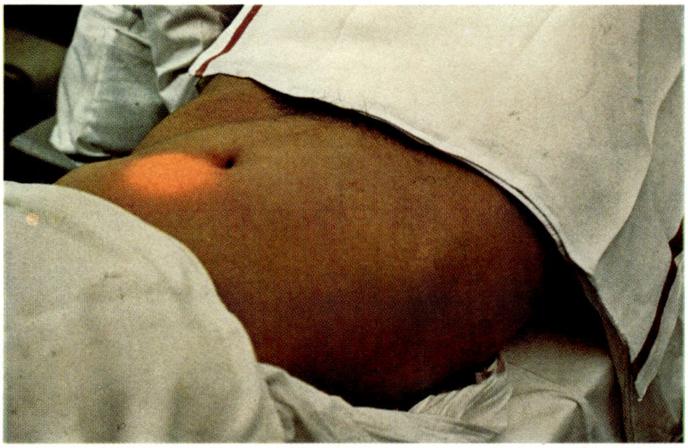

Fig. 63.
The lighted tip of the coloscope has reached the subcostal region, i.e. the splenic flexure of the colon.

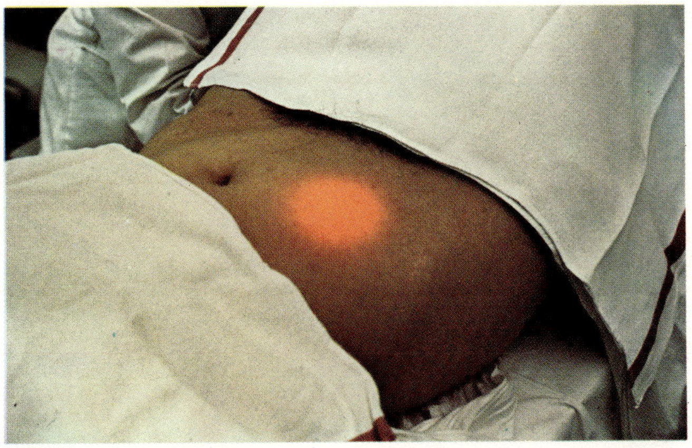

Fig. 64.
The epigastric region is lighted. The coloscope is in the transverse colon.

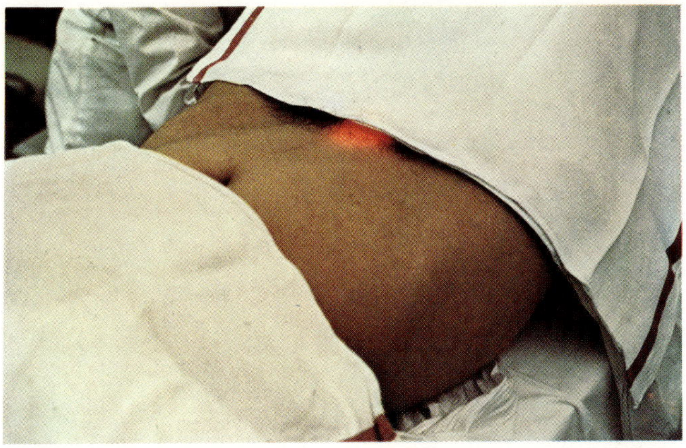

Fig. 65.
The lighted tip of the coloscope is in the right iliac fossa, i.e. in the cecum.

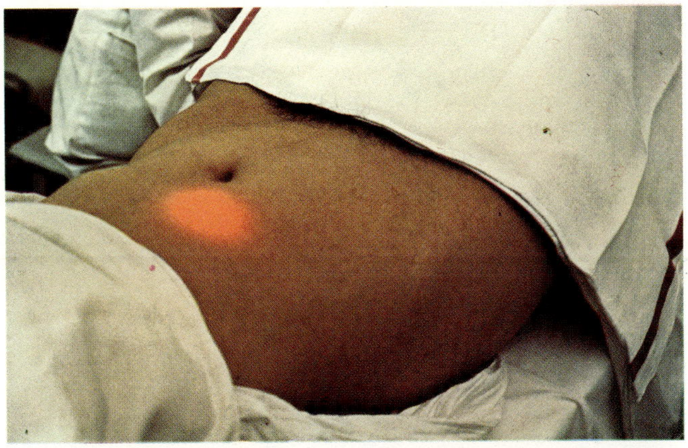

Fig. 66.
Transillumination of the suprapubic region. The coloscope is in the ileum.

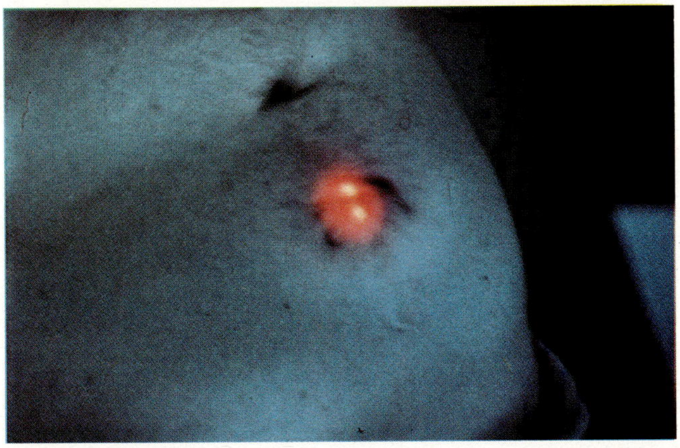

Fig. 67.
The lighted tip is impacted in the transverse colon, where the large bowel is virtually in direct contact with the wall of the abdomen on account of the presence of a laparocele. In this case examination was discontinued because the arrest of the tip of the coloscope at the laparocele would have made it hazardous to continue the examination.

coloscopy through colostomies

The same basic technique is used to examine the oral and aboral segments of the large bowel in patients with colostomies.

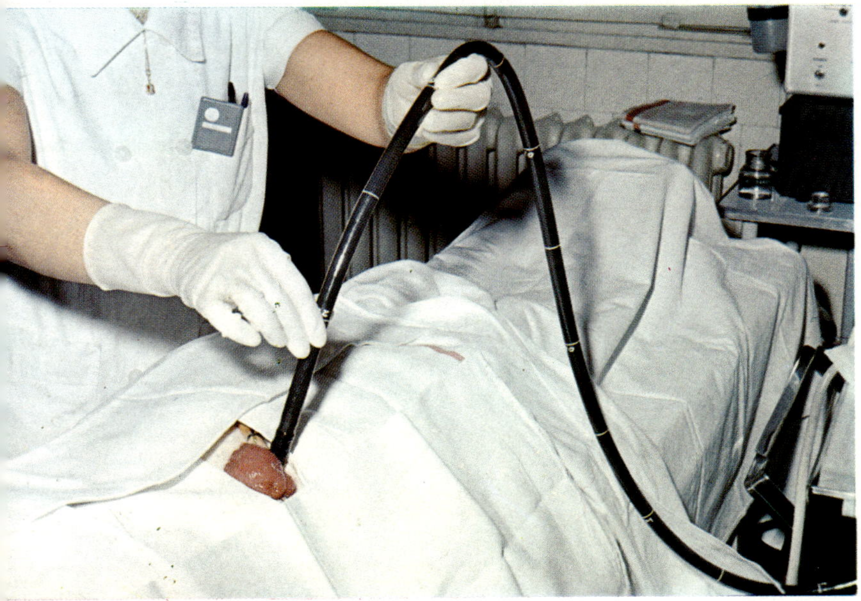

Fig. 68.
Coloscopy through a colostomy.
The patient is in the supine position.
The coloscope is inserted through a colostomy on the transverse colon.

Indications for coloscopy

1) Assessment of abnormal areas on barium enema.
2) Complete incontinence on barium enema.
3) Assessment of colic pain.
4) Unexplained abnormalities of the bowel function associated with myxorrhoea, occult bleeding or unexplained anaemia.
5) Assessment of inflammatory diseases of the colon.
6) Rectal bleeding (emergency coloscopy).
7) Inspection of colostomies.
8) Periodical check of surgical anastomoses.
9) Inspection of the colon in cases of rectal polyps.
0) Endoscopic polypectomy of the large bowel.

Contraindications

Causes of failure

bsolute contraindications

) Extreme debilitation.
) Toxic megacolon.
) Severe ulcerative proctocolitis.
) Diverticulitis.
) High grade diverticulosis with incomplete stenosis.
) Severe adnexitis and pelvic peritonitis.
) Recent surgery of the large bowel.
) Infectious diseases of the colon.

elative contraindications

) Incomplete stenosis.
) Radiation sigmoidocolitis.
) Severe respiratory insufficiency.
) Severe heart disease.

) Excessive tortuosity and angulation of the sigmoid colon.
) Dolichomegacolon.
) Presence of faeces due to imperfect bowel preparation.
) Adhesions due to previous abdominal surgery.
) Partial stenosis or adhesions due to periviscertis.

Complications

Perforation

Perforation can be caused by:

(a) direct impact of the coloscope against the wall of the large bowel;
(b) excessive insufflation of air;
(c) injury by the biopsy forceps;
(d) rupture of a diverticulum;
(e) during polypectomy.

The most serious complication of coloscopy is perforation. Perforation may be due to rupture of the bowel wall under the impact of the tip of the coloscope or on account of the excessive insufflation of air. The mucosa is distended and becomes pale. The insufflated air rebounds from the bowel wall. If the coloscope is now inserted further the bowel wall may be torn. Any complaint of pain by the patient should receive immediate attention. Fluoroscopy must be performed at once and should distension of the colon be observed the coloscope must be withdrawn until the pain subsides. Should pain persist coloscopy must be interrupted and a surgeon consulted.

Perforation may also occur during biopsy especially when the mucosa is brittle, or when there is a bleeding growth or even when the mucosa is normal but the forceps is inserted deeply. Biopsy is even more hazardous when performed on a distended colon.

The presence of diverticula considerably increases the possibility of perforation because the orifice of a diverticulum can easily be mistaken for the bowel lumen. The most dangerous are intramural diverticula which are hard to discern clearly. The insufflation of excessive amounts of air or progression of the coloscope due to manipulation by the endoscopist or to unexpected peristaltic contractions of the large bowel may cause the rupture of the diverticulum (Figs. 69 & 70).

Great caution is required in the presence of stricturing of the colon. In such cases the instrument should not be forced to avoid possible fissuration. The bowel lumen must measure not less than 1,5 cm to permit the passage of the tip of the fibrescope.

Bleeding

The other, less important, complication is bleeding. Excessive distension of the bowel wall can aggravate the bleeding in patients with ulcerative colitis. Distension of the lumen may enlarge the ulcerations and thus favour bleeding. Another factor is mechanical injury caused by the passage of the coloscope over the brittle mucosa of patients with ulcerative colitis.

Great caution must therefore be used in examining patients suffering from ulcerative proctocolitis, Crohn's disease or irradiation sigmoiditis.

Bleeding may be caused by injury of a neoplastic growth by the tip of the instrument or by the biopsy forceps. It is usually slight and negligible.

The possible complications of endoscopic polypectomy of the colon will be described separately.

Fig. 69.
Parietal diverticulum with
wide sac.

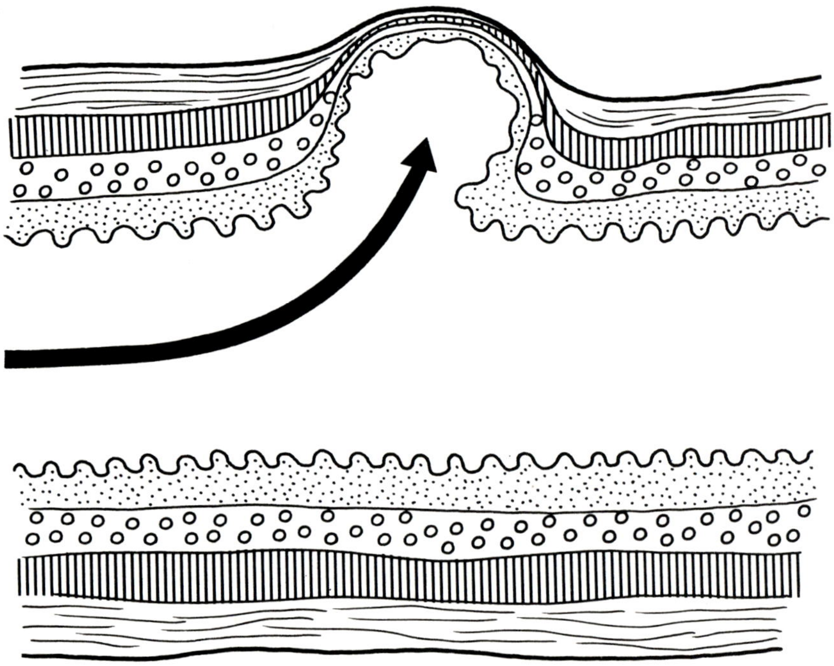

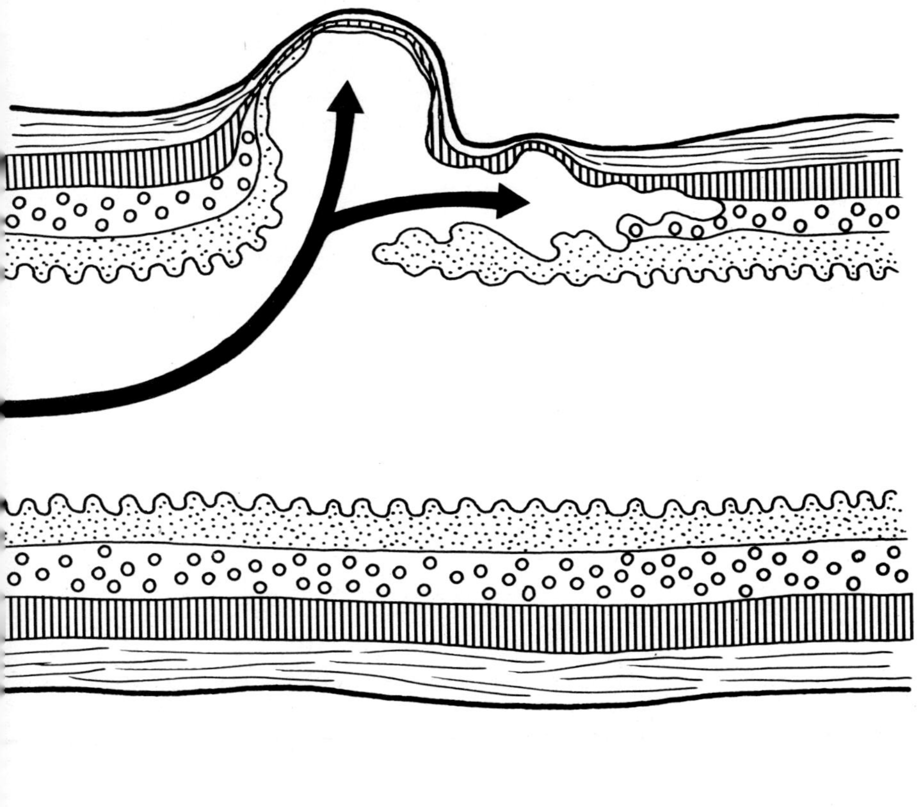

Fig. 70.
Parietal and intramural diverticulum. Part of the diverticular sac extends into the bowel wall.

Coloscopic form

Coloscopy n. ______

Family name First name Sex M

Date and place of birth

Addressing physician

Diagnostic problem

Barium enema (date)

Other diseases

Significant history data

Fibrescope model

Endoscopic findings

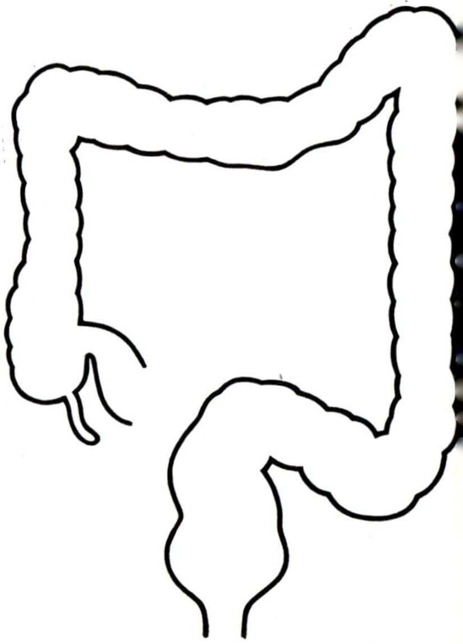

Endoscopic polypectomy

Diathermy snare type

Polyp retrieved by:

forceps or basket ☐
suction ☐
natural route ☐
lost ☐

Control after polypectomy

Complications

Notes

Diagnosis

		Sigmoid colon (1st portion)	Sigmoid colon (2nd portion)	Descending colon	Left flexure	Transverse colon	Right flexure	Ascending colon	Cecum	Ileo-cecal papilla	Ileum
	1) faeces	1	1	1	1	1	1	1	1	1	1
	2) blood	2	2	2	2	2	2	2	2	2	2
	3) mucus	3	3	3	3	3	3	3	3	3	3
	4) pus	4	4	4	4	4	4	4	4	4	4
gression the fibrescope	1) difficult	1	1	1	1	1	1	1	1	1	1
	2) bends	2	2	2	2	2	2	2	2	2	2
	3) stenosis	3	3	3	3	3	3	3	3	3	3
vel walls	1) distensible	1	1	1	1	1	1	1	1	1	1
	2) stiff	2	2	2	2	2	2	2	2	2	2
cosa (colour)	1) pink	1	1	1	1	1	1	1	1	1	1
	2) bright red	2	2	2	2	2	2	2	2	2	2
	3) pale	3	3	3	3	3	3	3	3	3	3
	4) brownish	4	4	4	4	4	4	4	4	4	4
pearance	1) smooth	1	1	1	1	1	1	1	1	1	1
	2) shiny	2	2	2	2	2	2	2	2	2	2
	3) dull	3	3	3	3	3	3	3	3	3	3
	4) granular	4	4	4	4	4	4	4	4	4	4
	5) oedematous	5	5	5	5	5	5	5	5	5	5
	6) brittle	6	6	6	6	6	6	6	6	6	6
	7) cobblestone	7	7	7	7	7	7	7	7	7	7
ustral folds	1) normal	1	1	1	1	1	1	1	1	1	1
	2) high	2	2	2	2	2	2	2	2	2	2
	3) pseudopolypoid	3	3	3	3	3	3	3	3	3	3
	4) low	4	4	4	4	4	4	4	4	4	4
	5) absent	5	5	5	5	5	5	5	5	5	5
od vessels	1) enhanced vascular networks	1	1	1	1	1	1	1	1	1	1
	2) vascular networks	2	2	2	2	2	2	2	2	2	2
	3) dilated veins	3	3	3	3	3	3	3	3	3	3
sions	1) skipped	1	1	1	1	1	1	1	1	1	1
	2) diffuse	2	2	2	2	2	2	2	2	2	2
	3) bleeding	3	3	3	3	3	3	3	3	3	3
cers	1) linear	1	1	1	1	1	1	1	1	1	1
	2) serpiginous	2	2	2	2	2	2	2	2	2	2
	3) punched-out	3	3	3	3	3	3	3	3	3	3
	4) furrow-like	4	4	4	4	4	4	4	4	4	4
	5) aphtha-like	5	5	5	5	5	5	5	5	5	5
	6) multiple, bleeding	6	6	6	6	6	6	6	6	6	6
	7) deep	7	7	7	7	7	7	7	7	7	7
	8) shallow	8	8	8	8	8	8	8	8	8	8
udopolyps	1) multiple	1	1	1	1	1	1	1	1	1	1
	2) diffuse	2	2	2	2	2	2	2	2	2	2
	3) brittle	3	3	3	3	3	3	3	3	3	3
	4) hard	4	4	4	4	4	4	4	4	4	4
ips	1) single	1	1	1	1	1	1	1	1	1	1
	2) multiple	2	2	2	2	2	2	2	2	2	2
	3) diffuse	3	3	3	3	3	3	3	3	3	3
Size	1) 0 - 5 mm.	1	1	1	1	1	1	1	1	1	1
	2) 0,5 - 1 cm.	2	2	2	2	2	2	2	2	2	2
	3) 2 - 4 cm.	3	3	3	3	3	3	3	3	3	3
	4) 5 cm.	4	4	4	4	4	4	4	4	4	4
Type	1) sessile	1	1	1	1	1	1	1	1	1	1
	2) pedunculated	2	2	2	2	2	2	2	2	2	2
Surface	1) smooth	1	1	1	1	1	1	1	1	1	1
	2) papillary	2	2	2	2	2	2	2	2	2	2
	3) lobulated	3	3	3	3	3	3	3	3	3	3
	4) ulcerated	4	4	4	4	4	4	4	4	4	4
	5) brittle	5	5	5	5	5	5	5	5	5	5
	6) hard	6	6	6	6	6	6	6	6	6	6
	7) infiltrated base	7	7	7	7	7	7	7	7	7	7
verticula	1) single	1	1	1	1	1	1	1	1	1	1
	2) multiple	2	2	2	2	2	2	2	2	2	2
	3) diffuse	3	3	3	3	3	3	3	3	3	3
	4) ulcerated	4	4	4	4	4	4	4	4	4	4
rcinoma	1) fungating	1	1	1	1	1	1	1	1	1	1
	2) ulcerated	2	2	2	2	2	2	2	2	2	2
	3) infiltrating	3	3	3	3	3	3	3	3	3	3
	4) stenosing	4	4	4	4	4	4	4	4	4	4
	5) bleeding	5	5	5	5	5	5	5	5	5	5
rgical anastomoses	1) soft	1	1	1	1	1	1	1	1	1	1
	2) patent	2	2	2	2	2	2	2	2	2	2
	3) stenosed	3	3	3	3	3	3	3	3	3	3
	4) recurrence of cancer	4	4	4	4	4	4	4	4	4	4
	5) suture remnants	5	5	5	5	5	5	5	5	5	5
psy											
oto											
e-film											
rium enema filed											
stologic specimen N.											
rgical findings											

The normal colon

he end portion of the alimentary canal, the large intestine, easures about 180 cm in length and from 3 to 8 cm in ameter. Its walls vary in thickness from 0.5 to 2 cm.
he sigmoid colon, about 4-5 cm in diameter, interposed etween the rectum and the descending colon, is 15 to 80 cm ng, and forms wide convolutions of variable aspect. It is stinguished endoscopically by the presence of rather ell-marked semilunar folds and by the velvety appearance of s mucosa.
he sigmoid colon communicates with the descending colon ith which it forms a more or less acute angle igmoid-descending colon junction). The descending colon as the appearance of a long, straight tube marked by circular austrations which are lower but more regular in shape and aced at more regular intervals than those of the sigmoid olon. The internal appearance of the large bowel reproduces e inverse of the swellings and incisures on its outer surface though this aspect is not static but changes continually under e effect of the contractions of the intestinal musculature. The ucosa of the descending colon is smooth, shiny, thin and eyish-pink in colour.
t the proximal extremity of the descending colon the left or olenic flexure can be recognised by its dome-like appearance.
he transverse colon is easily distinguishable on account of its aracteristic triangular haustrations.
is lined with greyish-pink, smooth, shiny mucosa through hich the underlying fine vascular network shows up.
he hepatic flexure has a characteristic pagoda-like appearance ue to the arrangement of its haustral folds.
he ascending colon is larger than the left colon and is lined ith a mucosa forming semicircular folds.
he cecum has the appearance of a large cul-de-sac, crossed by rge folds which divide it up into pouch-like recesses.
he cecum contains the orifice of the appendix. It is ound-shaped and hidden under a mucosal fold. The orifice f the ileum is situated on the wall of the ascending colon, bout 10 cm from the cecum. Il has a circular, bilabial or oval hape and measures from 1.5 to 3 cm in diameter. It is tuated on the tip of a conical protrusion which projects bout 1 cm into the large bowel. The orifice functions like a alve which opens when a peristaltic contraction approaches it n account of a gastro-ileal reflex mechanism. The large testine can be divided into two sections—the right colon, xtending from the cecum to the middle third of the transverse olon inclusive, supplied by the superior mesenteric artery— nd the left colon, extending down to the pelvic colon nclusive, supplied by the inferior mesenteric artery. The veins rain into the portal vein. The lymphatics drain into the uperior mesenteric, retropancreatic and juxta-aortic lymph odes.

Fig. 71.
Endoscopic view of the first portion of the sigmoid colon.

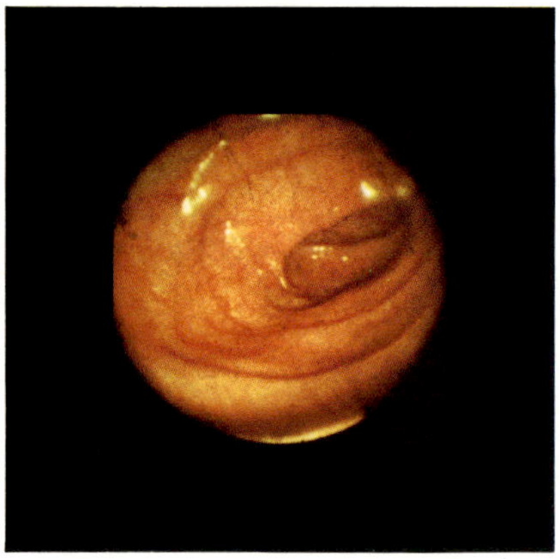

Fig. 72.
Middle portion of the sigmoid colon. The mucosa is smooth, shiny, and forms semicircular folds.

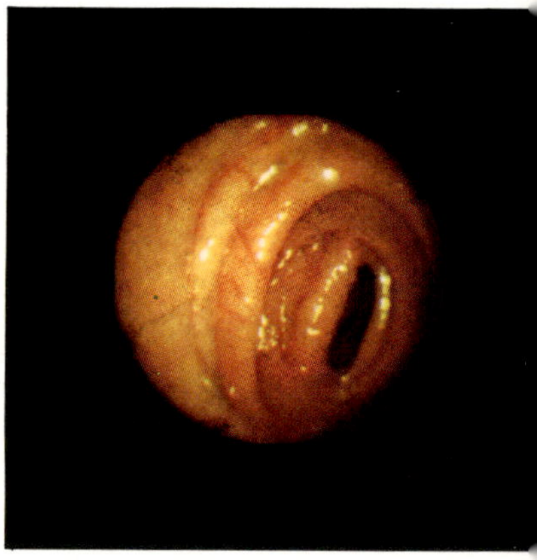

g. 73.
ıgulation of the distal portion
the sigmoid colon at its
ınction with the descending colon.

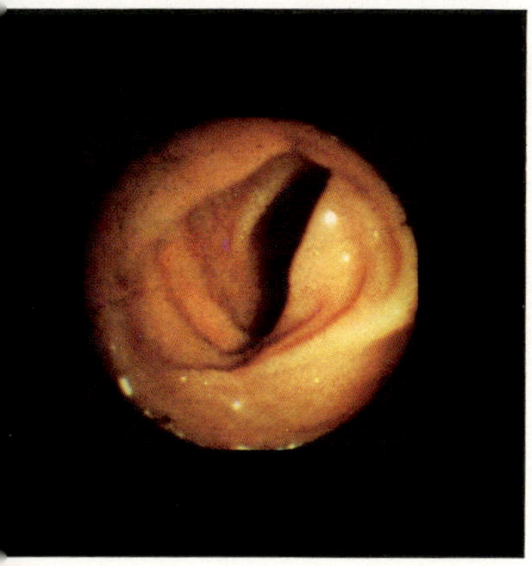

Fig. 74.
General view of the descending portion of the sigmoid colon. Semicircular folds.

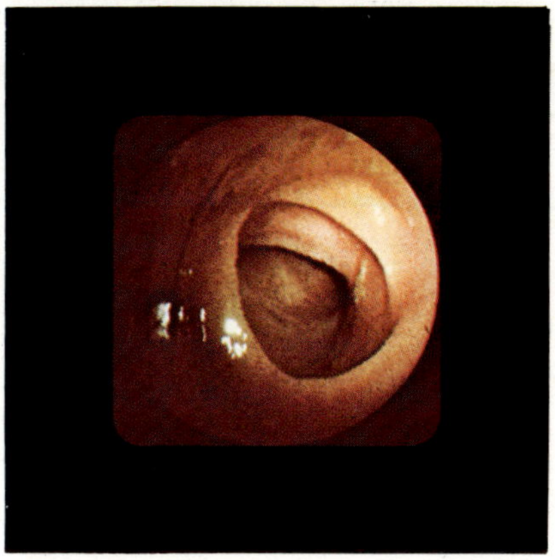

g. 75.
ɔwer third of the descending
lon just above the
ɡmoid-descending colon
ınction.

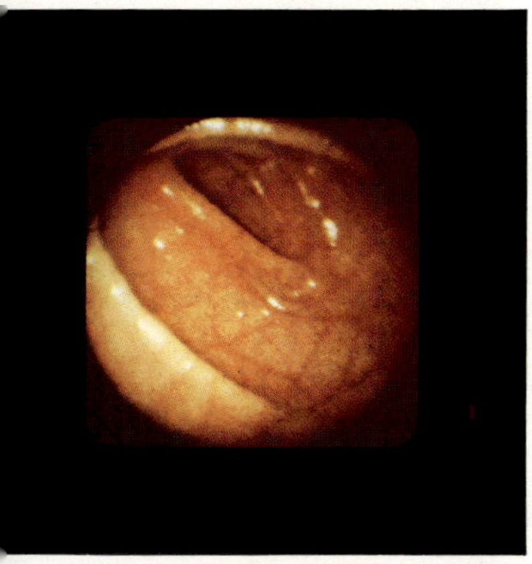

Fig. 76.
General view of the descending colon. The bowel forms a tunnel with circular haustrations.

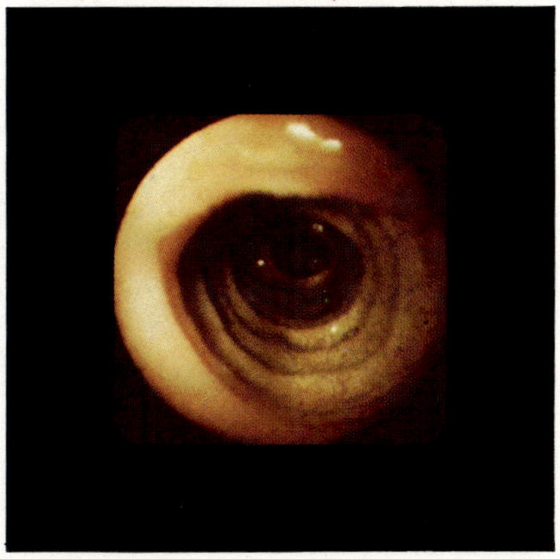

Fig. 77.
Splenic flexure of the large bowel.

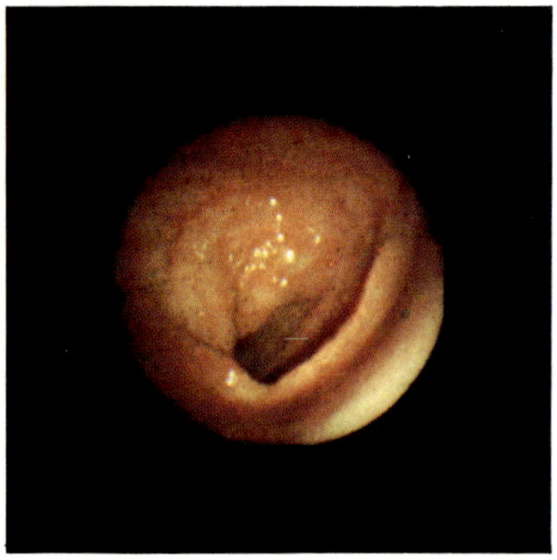

Fig. 78.
Close-up view of the splenic flexure.

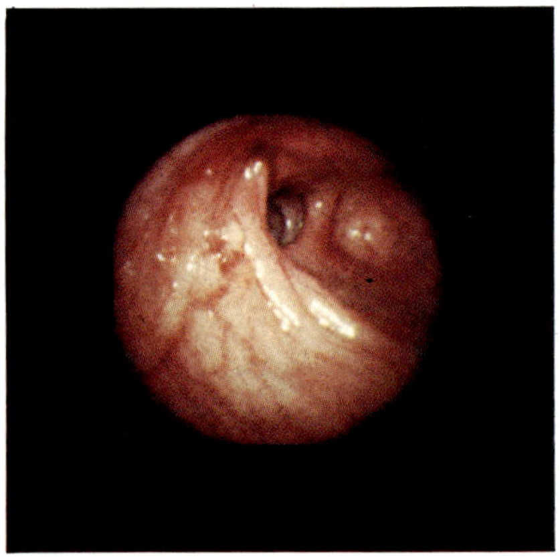

Fig. 79.
Descending colon and left flexure. The insufflation of a large volume of air has distended the bowel wall excessively and has altered the normal circular shape of the haustral folds and the endoscopic appearance of the mucosa. besides narrowing the splenic flexure.

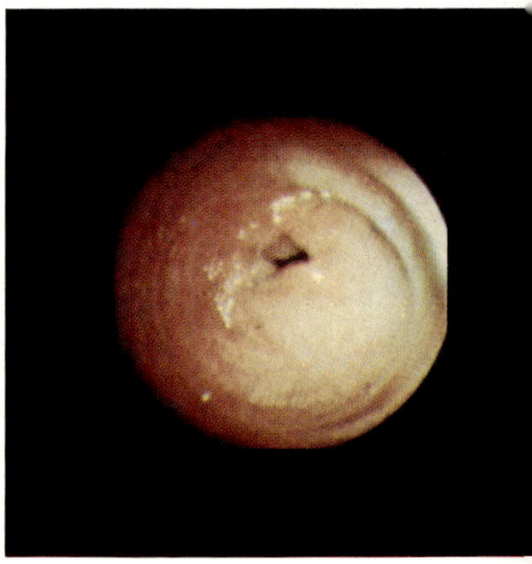

Fig. 80.
General view of the transverse colon. Note the characteristic triangular folds. Routine illumination (150 W).

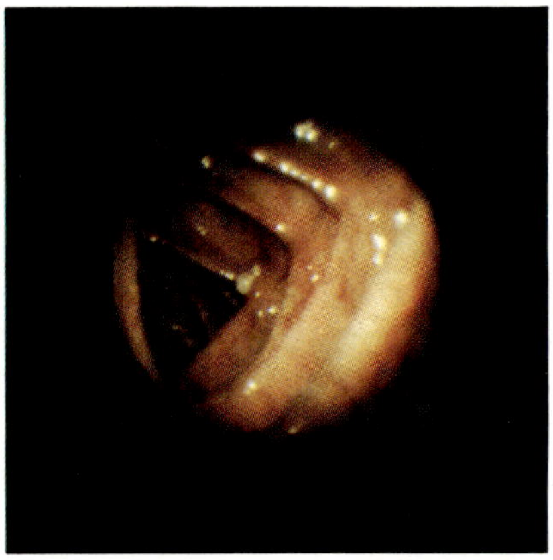

Fig. 81.
Same view with 300 W illumination.

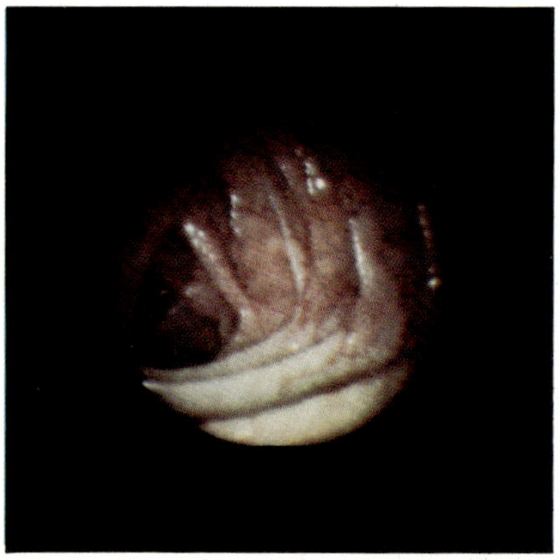

Fig. 82.
Middle third of the transverse colon

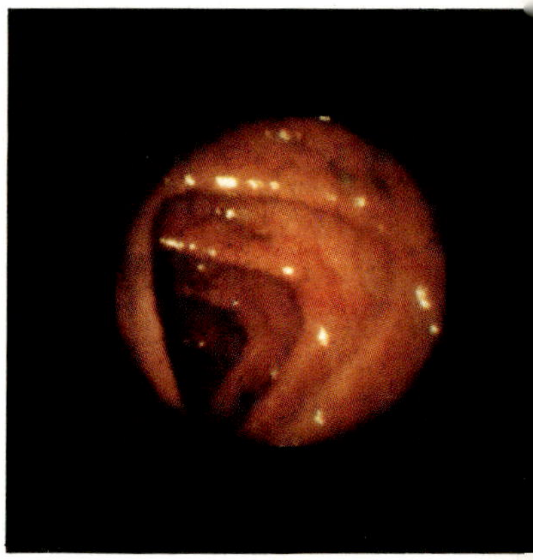

g. 83.
rst portion of the transverse
lon. The triangular folds can
st be made out.

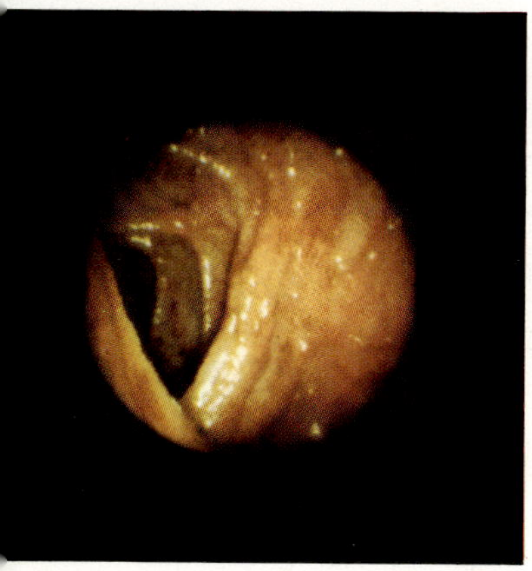

Fig. 84.
Close-up view of the wall of
the transverse colon.

g. 85.
etail of the middle third of the
ansverse colon, a possible blind
ot.

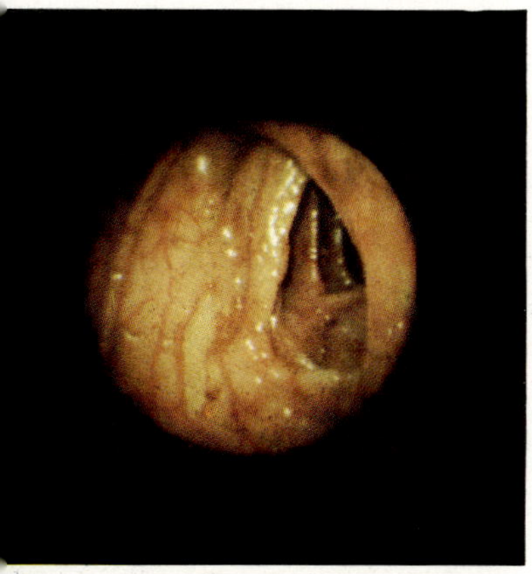

Fig. 86.
Transverse colon with partial
visualization of the hepatic flexure.

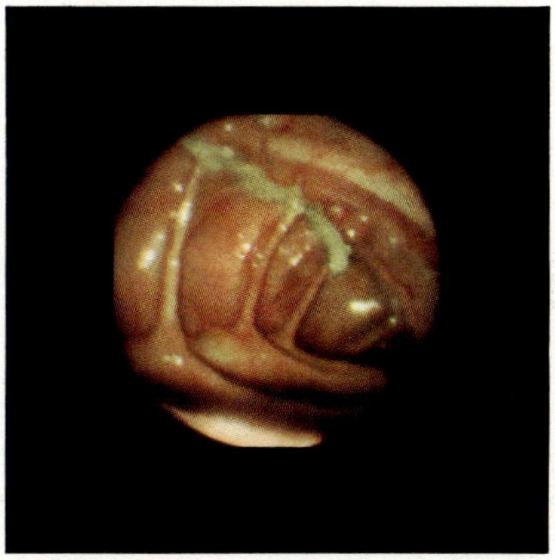

Fig. 87.
Close-up view of the same case.

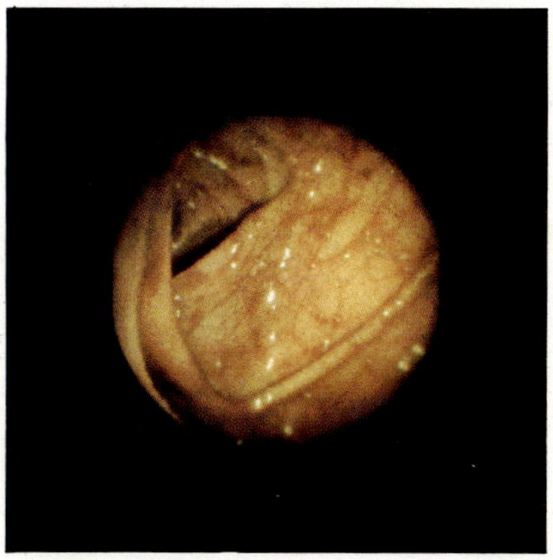

Fig. 88.
Hepatic flexure of the colon. Characteristic pagoda-like arrangement of the folds.

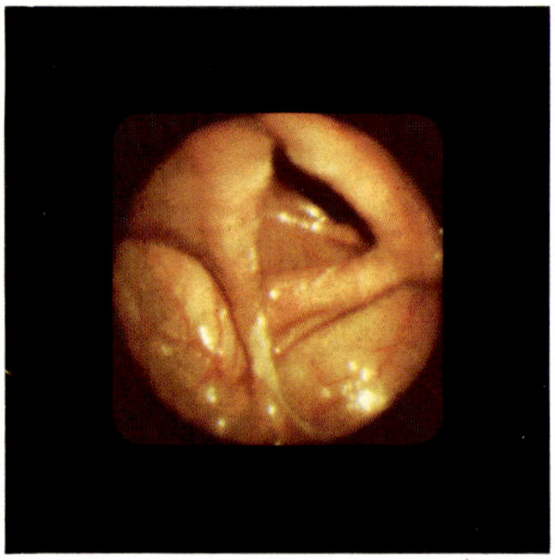

Fig. 89.
Persistent contracture of the hepatic flexure.

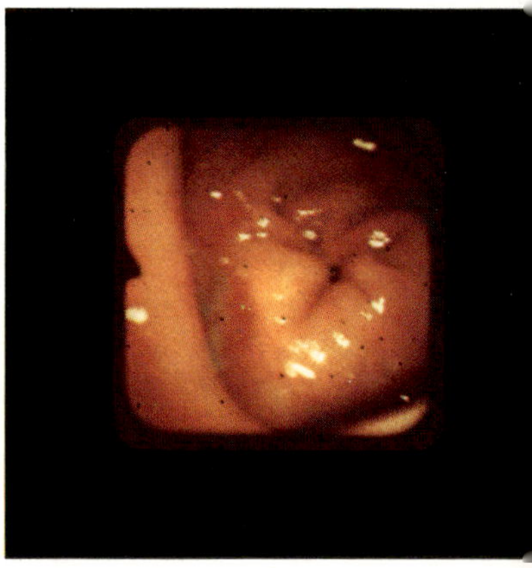

Fig. 90.
Ascending colon. Without excessive distension the bowel has an almost triangular form.

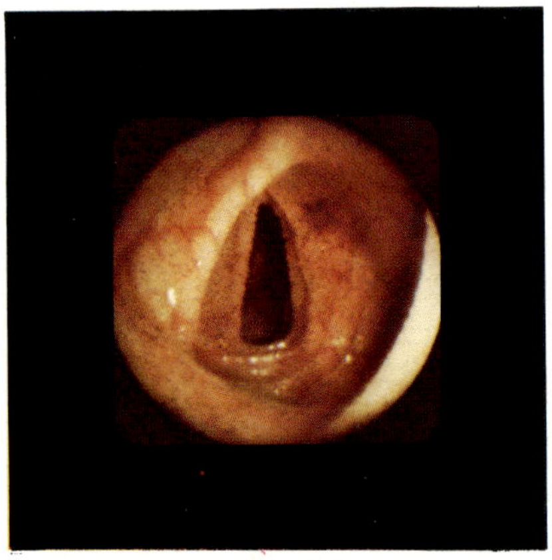

Fig. 91.
Same portion of the ascending colon after insufflation. The bowel wall is distended. The folds are semicircular.

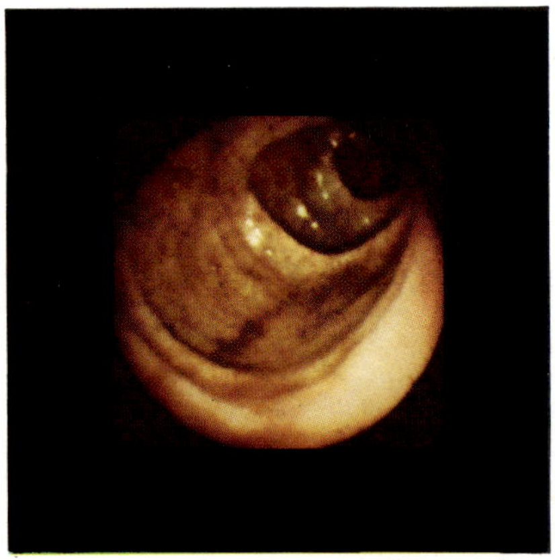

Fig. 92.
Retrograde view of the ascending colon. The tip of the fibrescope inspects the lumen from down up.

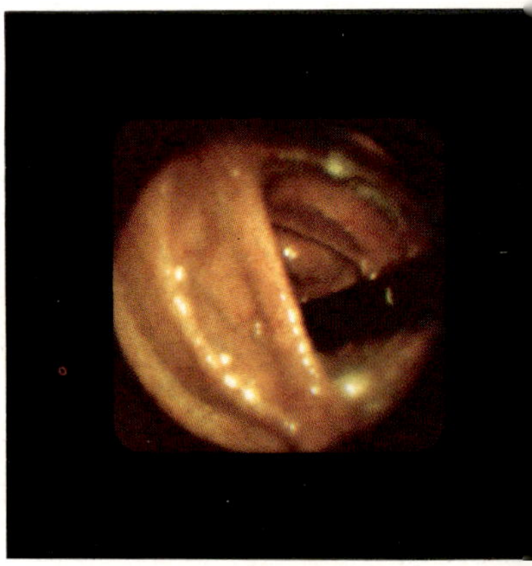

ig. 93.
ndoscopic view of the ascending
olon. The folds are semicircular.

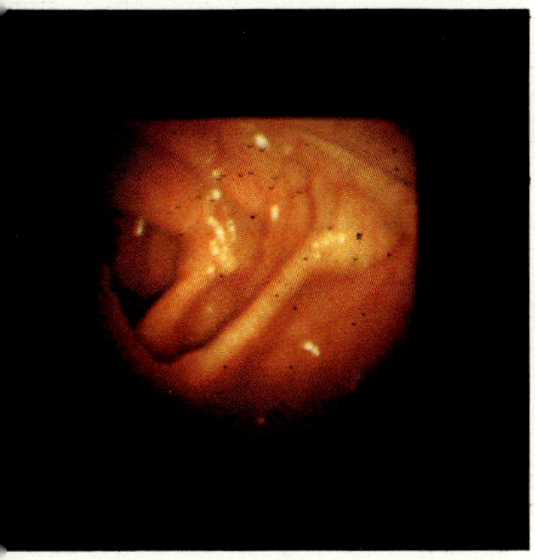

Fig. 94.
Ascending colon. Close-up view of the semicircular folds.

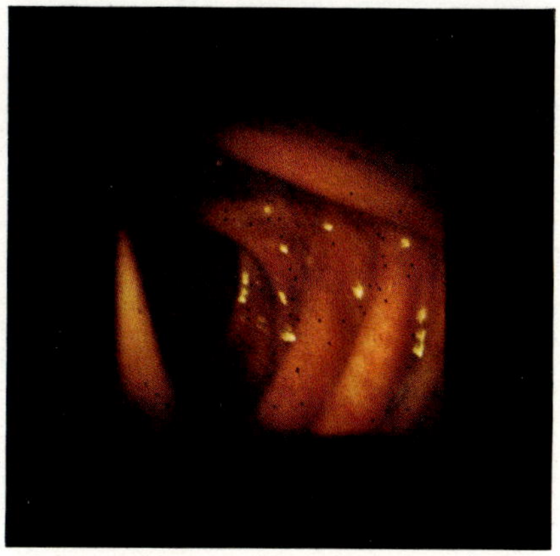

ig. 95.
ndoscopic of the ascending
olon distended by excessive
nsufflation.

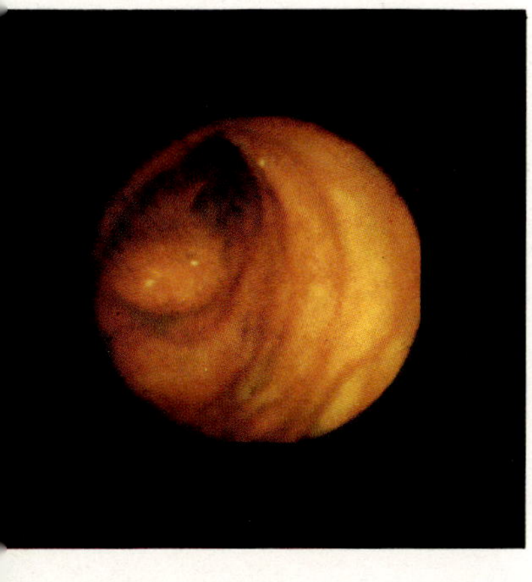

Fig. 96.
Same case. Retrograde view of the ascending colon.

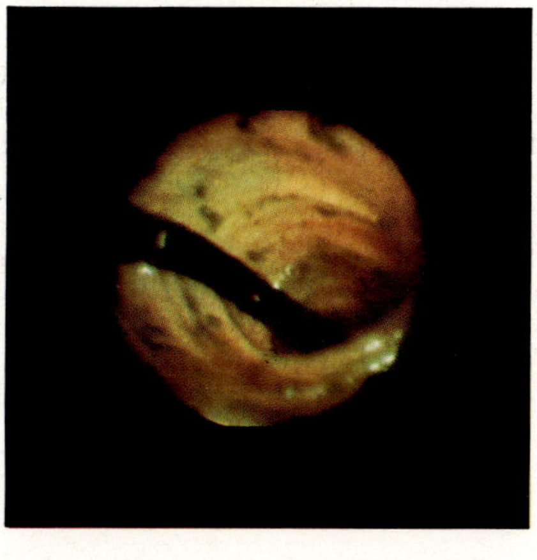

Fig. 97.
Endoscopic view of the fundus of the cecum. The folds converge towards the distal extremity forming characteristic pouches.

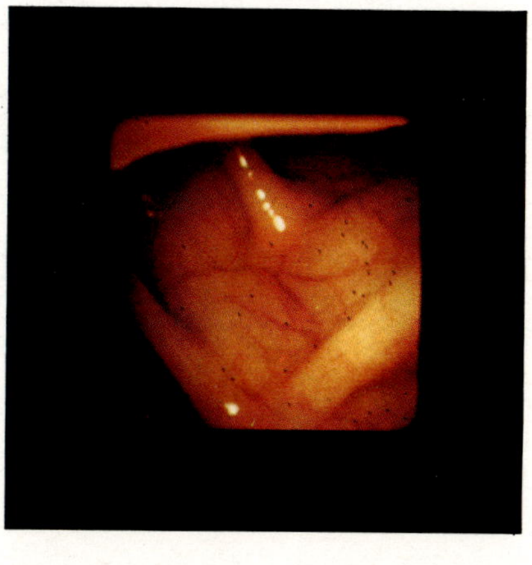

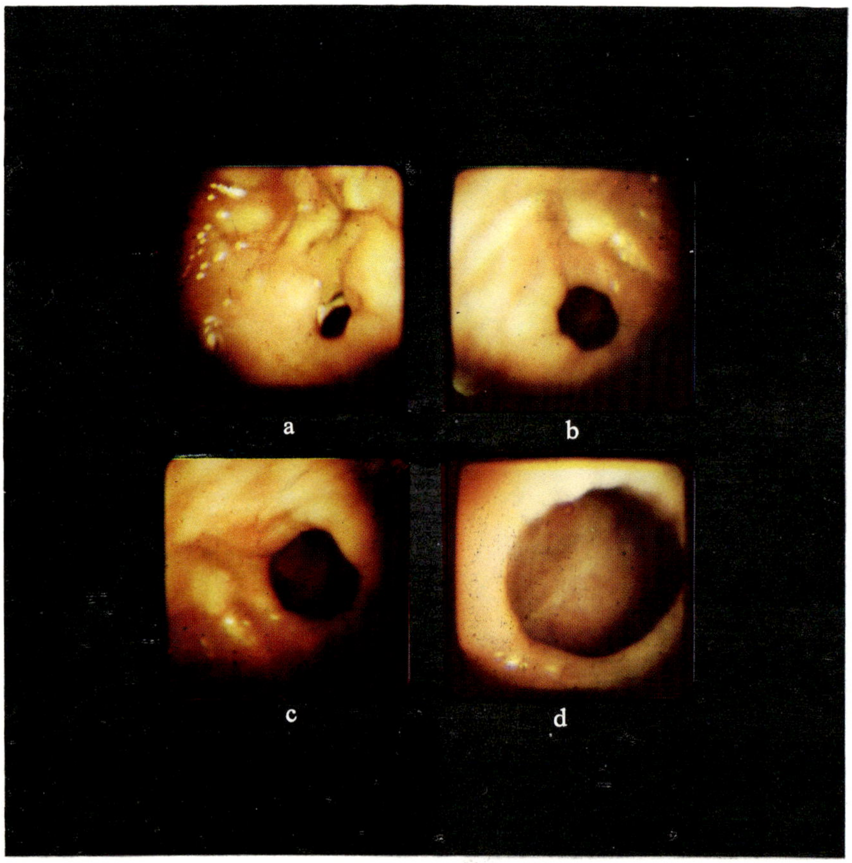

Fig. 98.
Different functional aspects of a round-shaped ileocecal papilla (Olympus CF-LB).

Fig. 99.

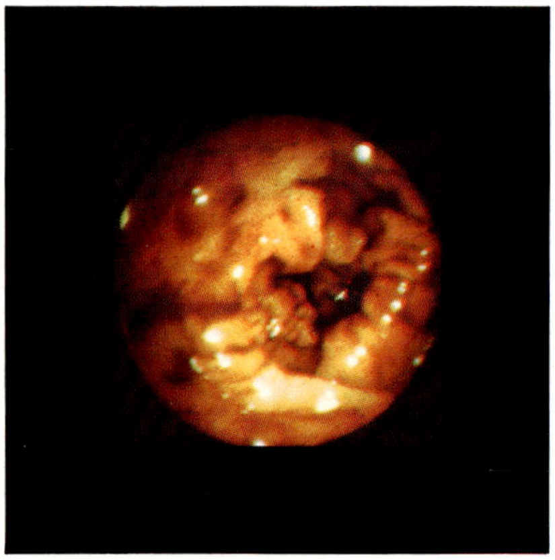

Fig. 100.

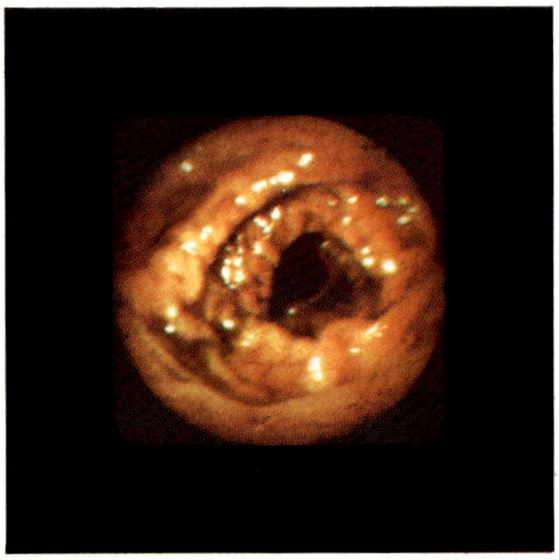

Figs. 99 & 100.
Ileocecal papilla.

ig. 101.

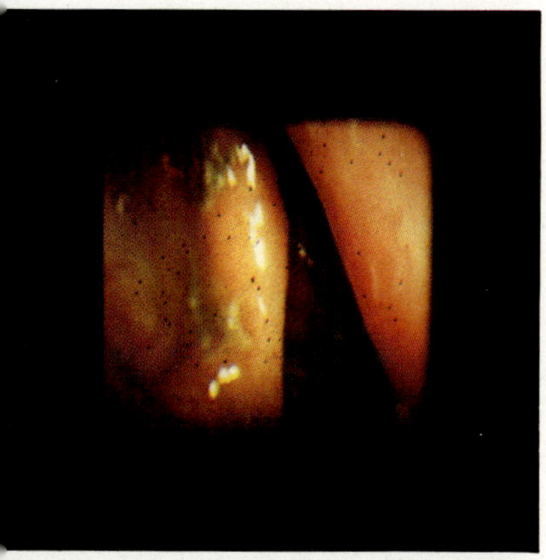

Fig. 102.

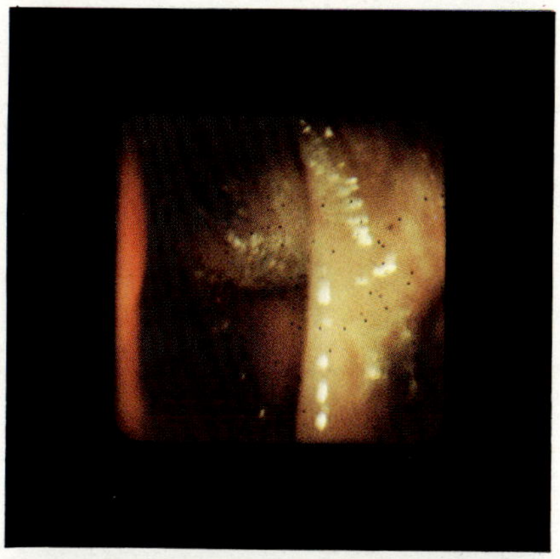

Figs. 101 & 102.
Ileocecal papilla of bilabial configuration. The terminal ileum is visible in Fig. 102.

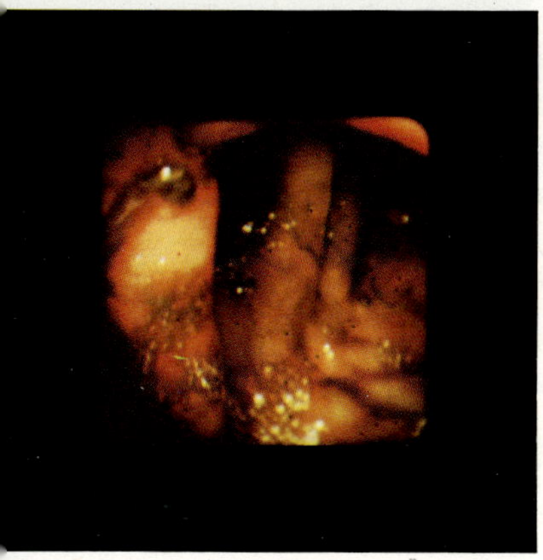

ig. 103.

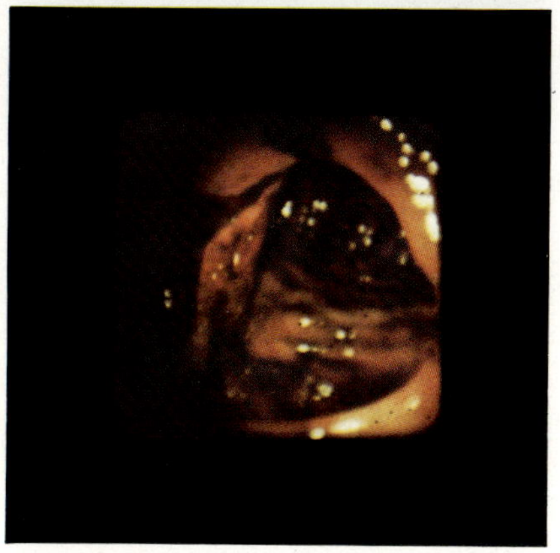

Fig. 104.

Figs. 103 & 104.
Endoscopic view of the last ileal loop.

ig. 105.

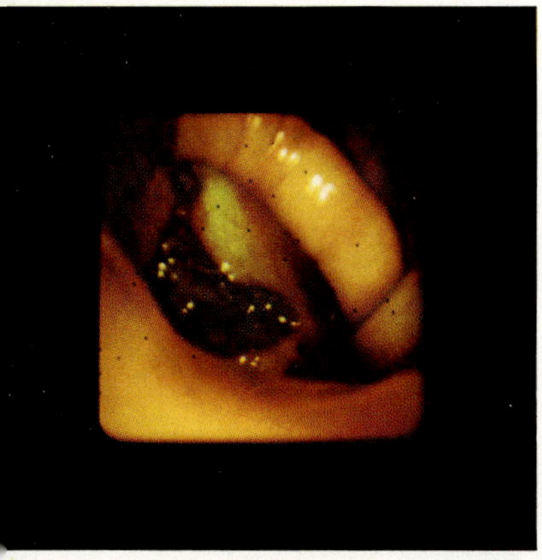

Fig. 106.

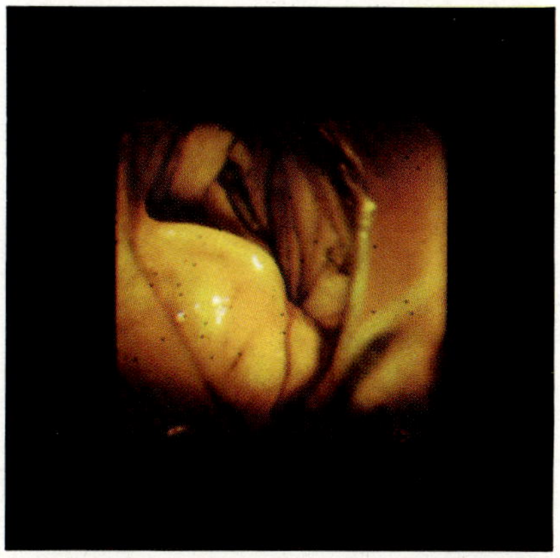

Fig. 105.
Terminal ileum.
Fig. 106.
Ileum.

Part two

Abnormal motor function of the colon (irritable colon)

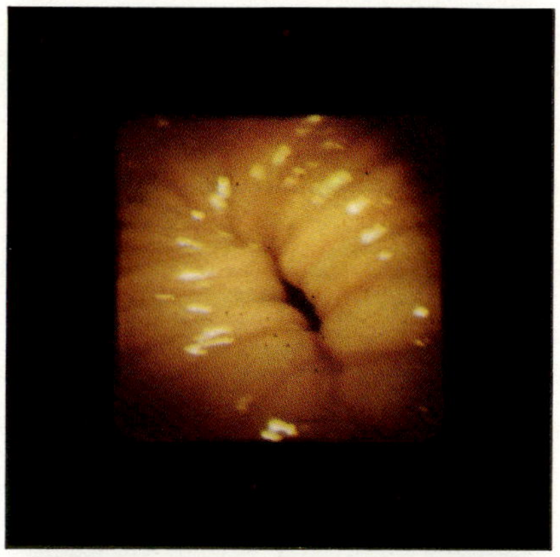

Fig. 107.
Contracture of the colon. Severe spasm at the middle third of the descending colon simulating a sphyncter and impeding further progression of the fibrescope. The mucosa is greyish-pink, smooth and shiny and forms numerous folds.

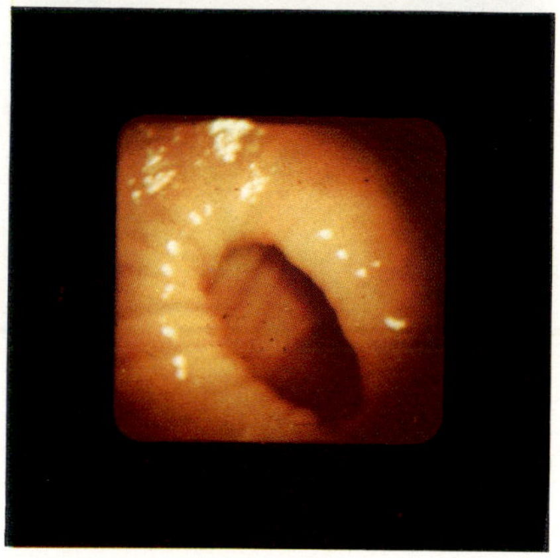

Fig. 108.
The spasm recedes spontaneously after a few minutes bringing into view the pink mucosa which lines the descending colon.

Fig. 109.
Serial photographs showing repeated spasms of the descending colon: *a* to *i* show the periodical contracture and distension of the colon.

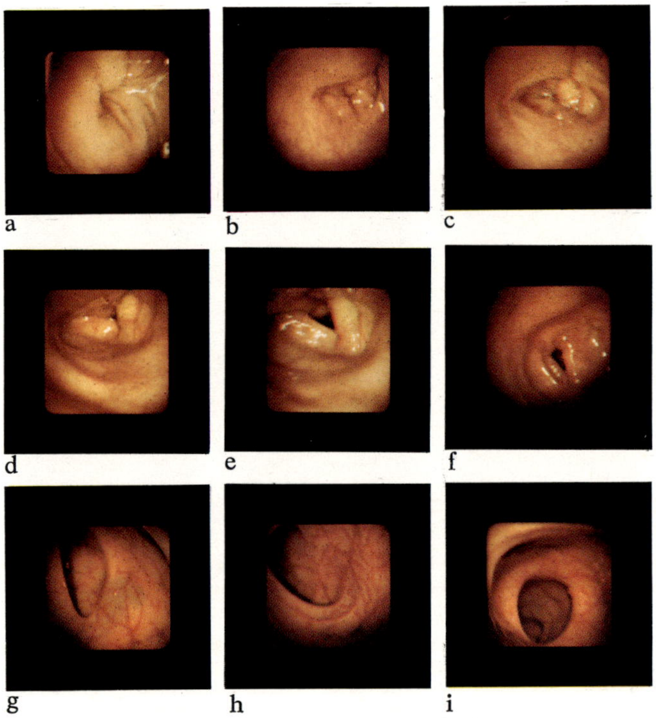

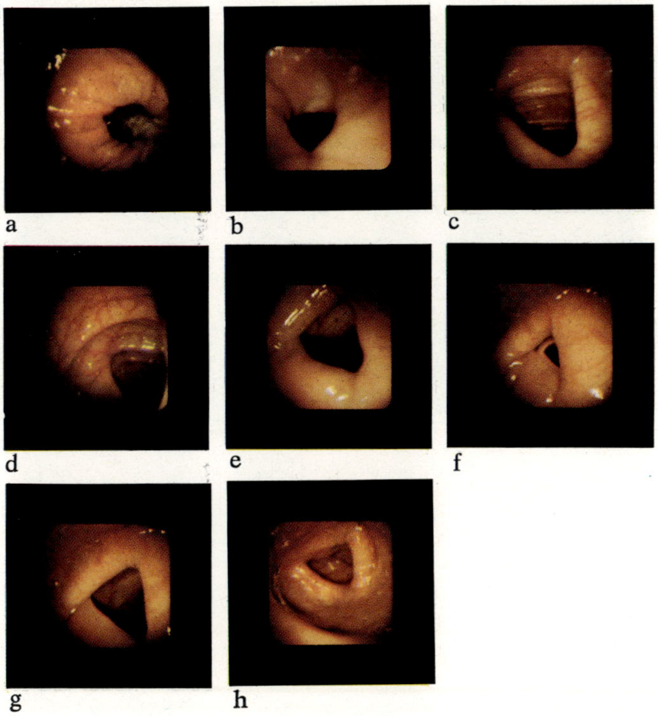

Fig. 110.
The motor disorder is also demonstrable in another part of the colon. The bowel wall stimulated by air insufflation and by the endoscope contracts and relaxes in a disorderly fashion. *a* shows a circular spasm which becomes more severe in *b*. The spasm recedes suddenly in *c* and *d*, only to recur with even greater intensity in *e* and *f* and to recede again in *g* and *h*.

Melanosis of the colon

Fig. 111.
Melanosis of the sigmoid colon.

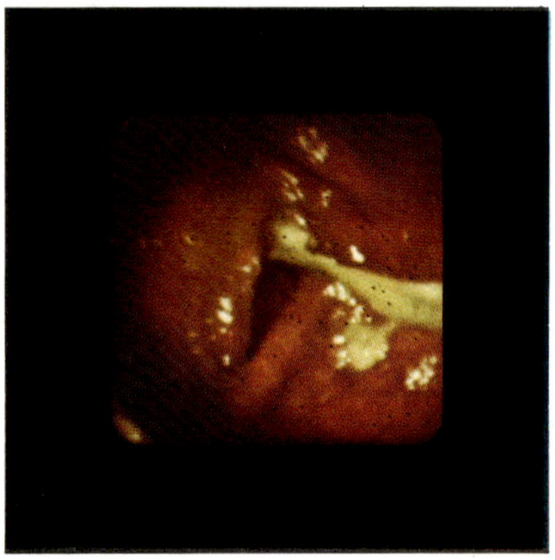

Fig. 112.
Melanosis of the descending colon.

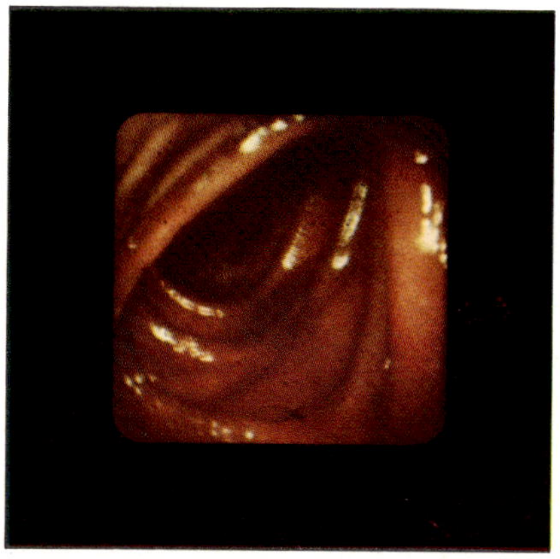

Fig. 113.
Melanosis of the splenic flexure.

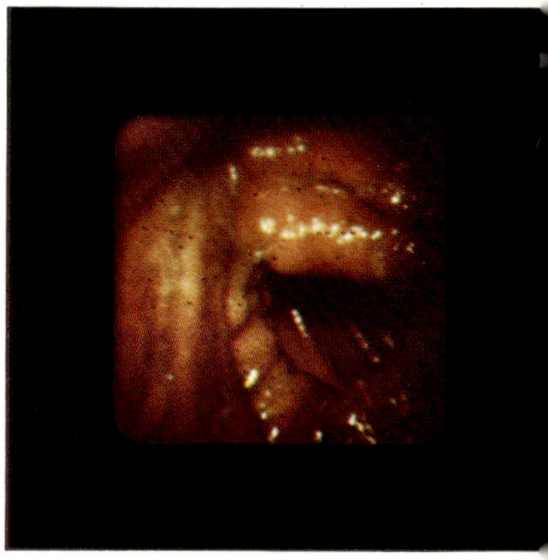

Fig. 117.
Early melanosis of the colon. Islets of brownish-grey mucosa in the colon.

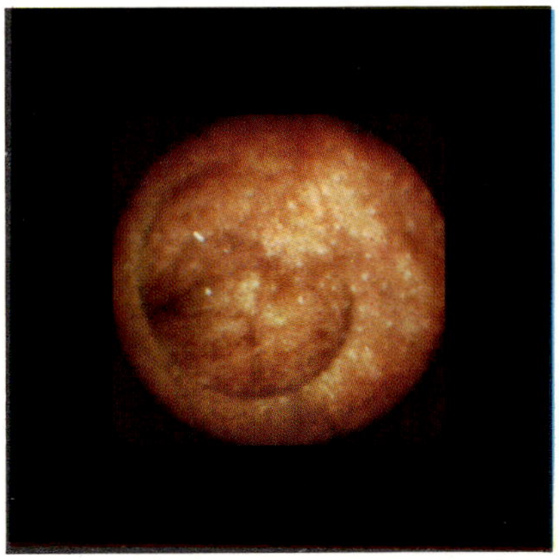

Fig. 118.
Melanosis of the sigmoid colon. Stripes of brownish-black mucosa separated by shiny yellowish striae in the sigmoid colon.

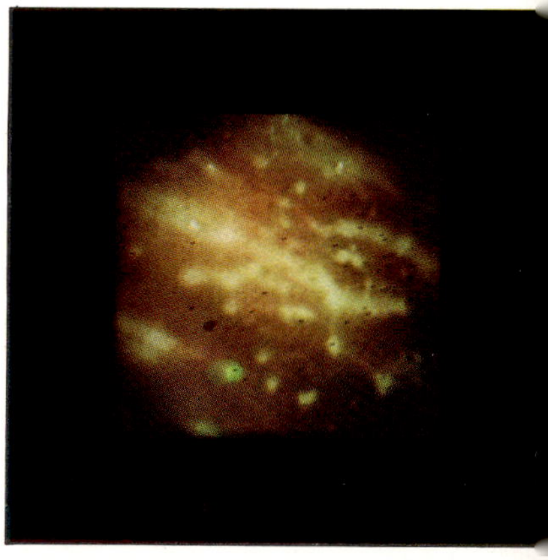

[T]he mucosa of the large intestine, from the rectum to the [c]ecum, is brown in colour and forms high, irregularly [o]edematous folds which are easily distensible by insufflation; [t]he typical brownish-black colouration of melanosis is also [p]resent on the ileocecal papilla and in the last loop of the [il]eum, as shown in the following series of endoscopic [p]hotographs taken in the same case.

[F]ig. 114.
[M]elanosis of the ascending colon.

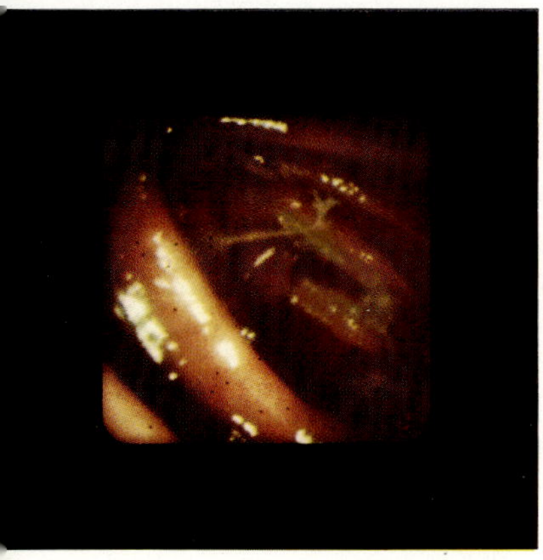

Fig. 115.
Melanosis of the ileocecal papilla.

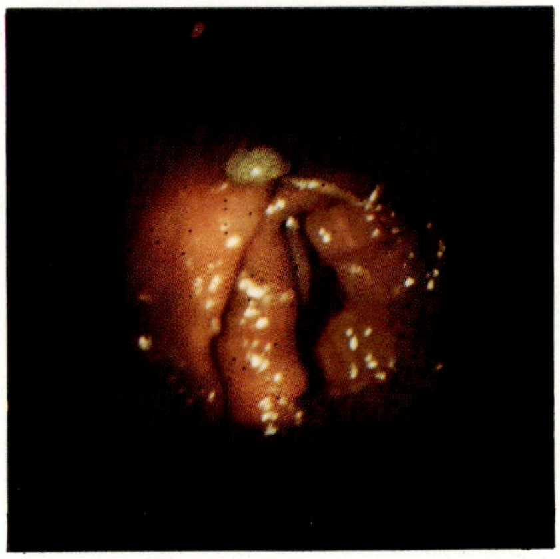

Fig. 116.
Melanosis of the last loop of the ileum.

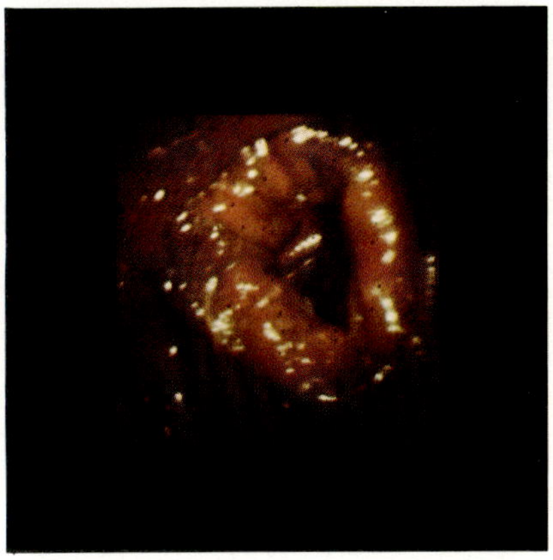

[F]ig. 119.
[C]oloscopic biopsy. Histologic [p]reparation. Melanosis of the colon. [I]nfiltration of the stroma with [r]eticulum cells.

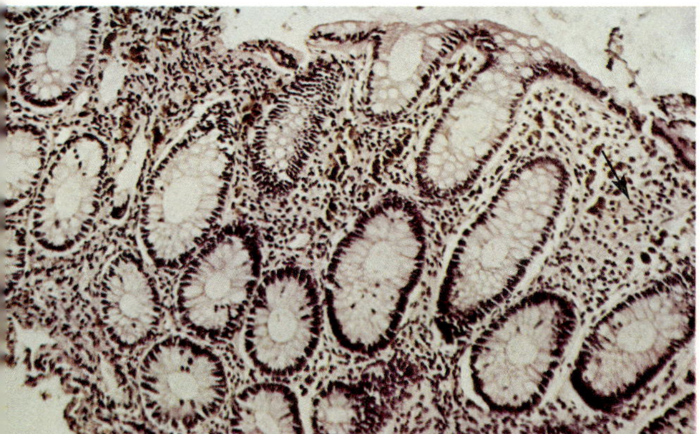

Fig. 120.
Detail of the same photomicrograph with higher magnification.

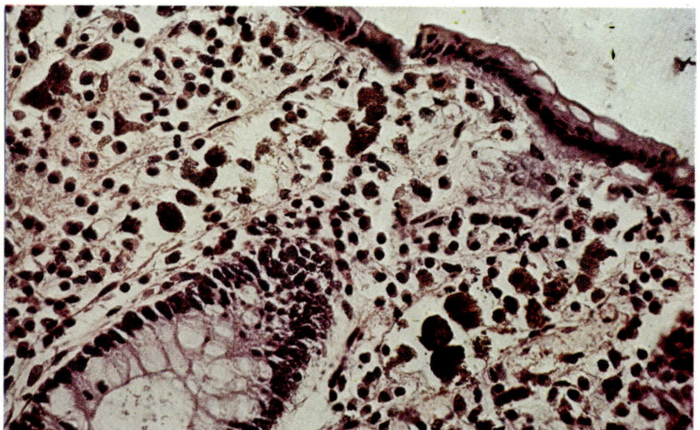

Diverticulosis of the large intestine

Fig. 121.
Barium enema.

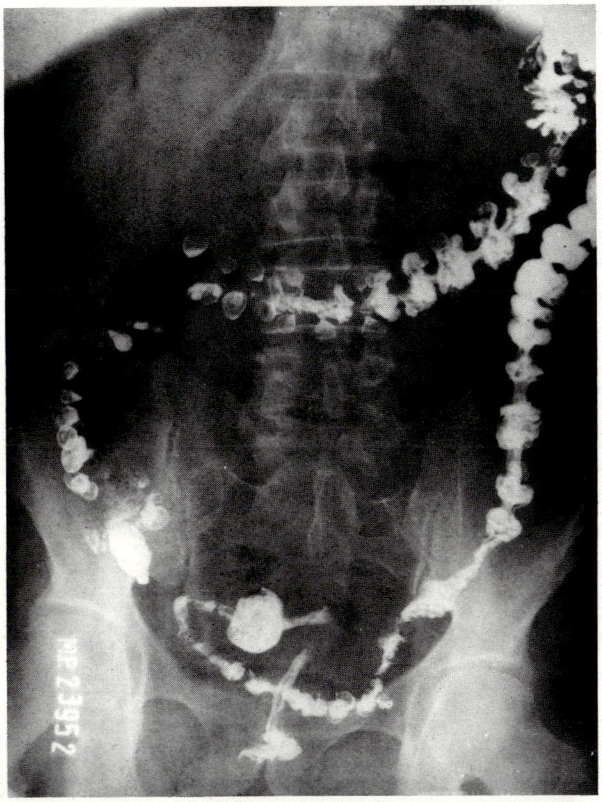

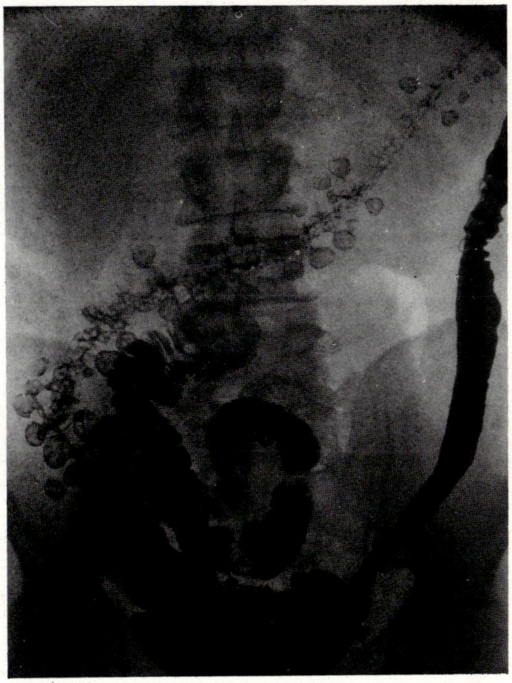

Fig. 122.
Barium enema after spontaneous evacuation.
The colon is hypertonic, intensely haustrated. Round marginal protuberances of various sizes, mostly pedunculated and completely independent, most numerous in the right and transverse colon.

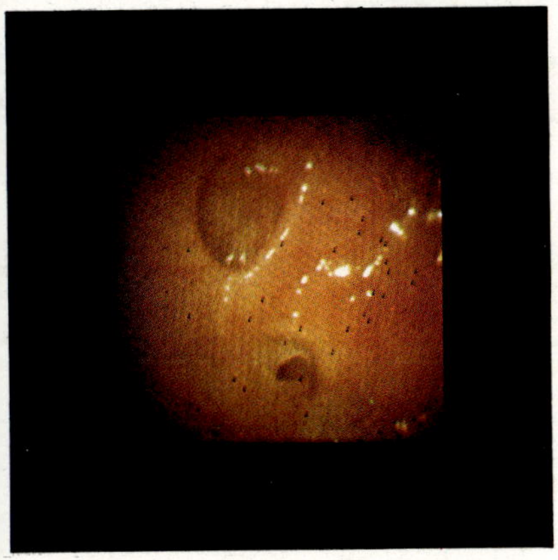

Fig. 123.

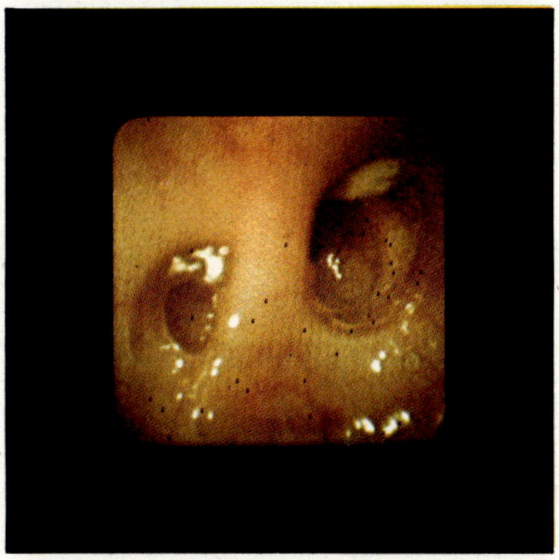

Fig. 124.

Figs: 123, 124 & 125.
Endoscopic examination of the colon reveals numerous diverticula which have the appearance of pouches of various shapes and sizes, with an open, oval orifice. At the hepatic flexure (Fig. 125) there is a very large diverticulum with a wide opening which can be mistaken for the bowel lumen. The diverticula are lined with smooth, shiny, greyish-pink mucosa with a normal vascular network. Some contain mucoserous material (Fig. 124).

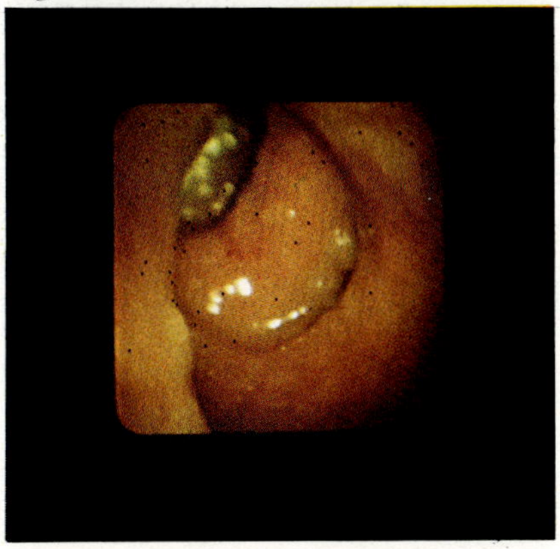

Fig. 125.

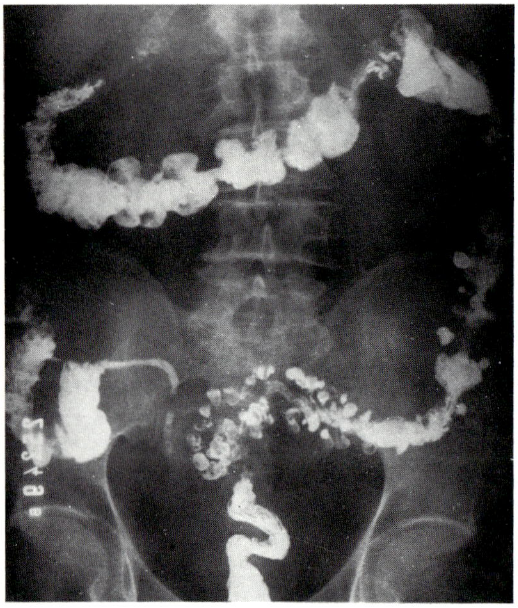

Fig. 126.
Barium enema.
Diverticulosis of the ileopelvic portion of the colon.

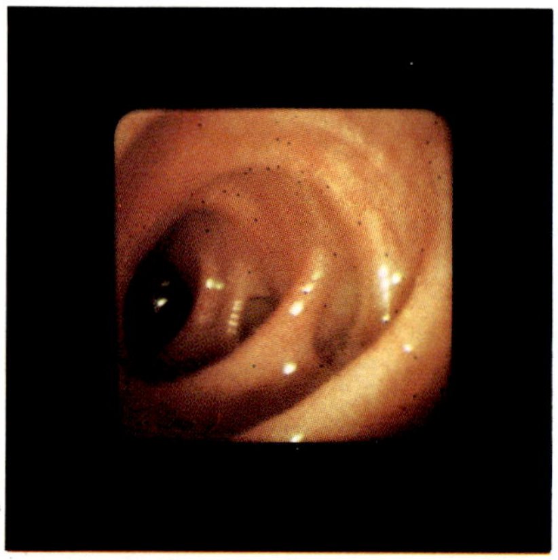

Fig. 127.
Diverticula of the lower third of the descending colon.

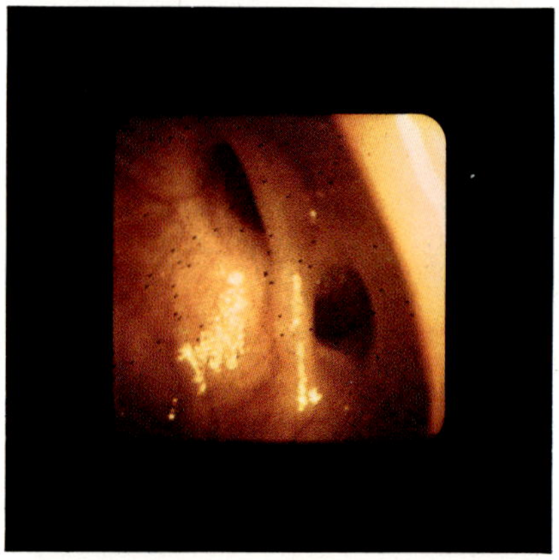

Fig. 128.
Same case. Close-up endoscopic view.

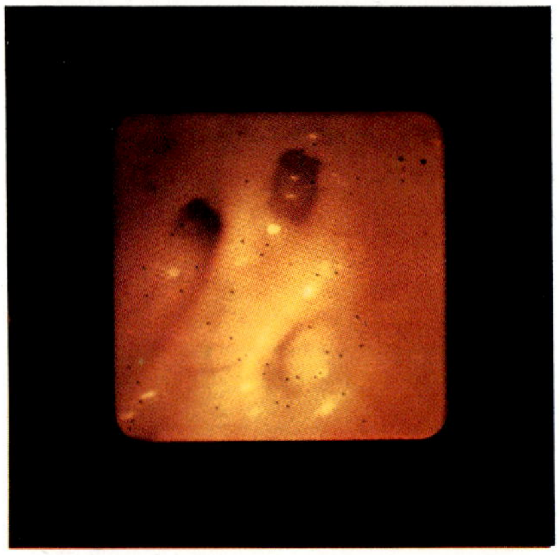

Fig. 129.
Multiple diverticulosis of the sigmoid colon.

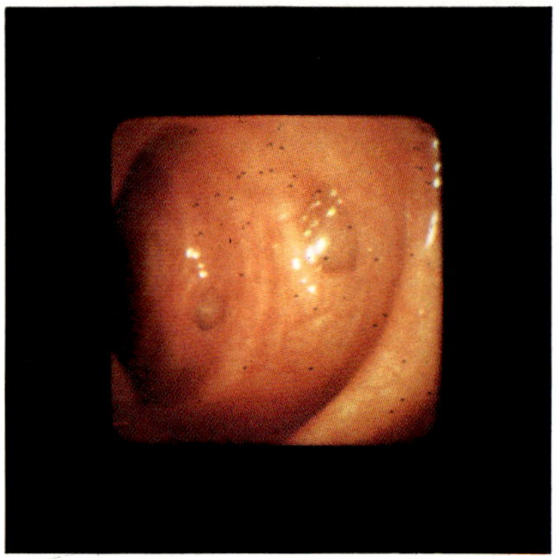

Fig. 130.
Shallow diverticula of the first portion of the sigmoid colon.

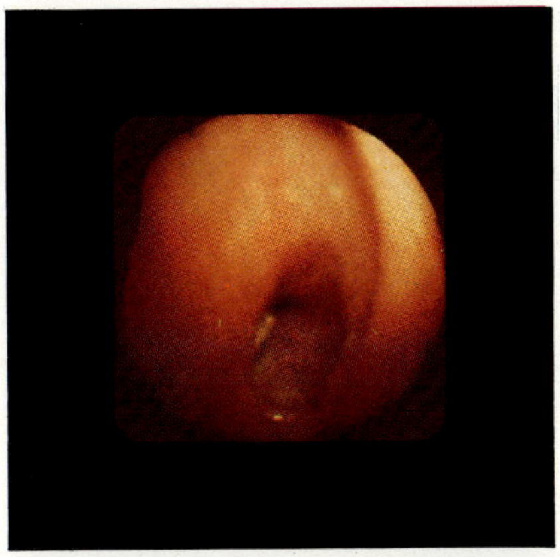

Fig. 131.
Large diverticulum of the sigmoid colon with oval orifice. The diverticulum is lined with hyperaemic mucosa and contains mucoserous material.

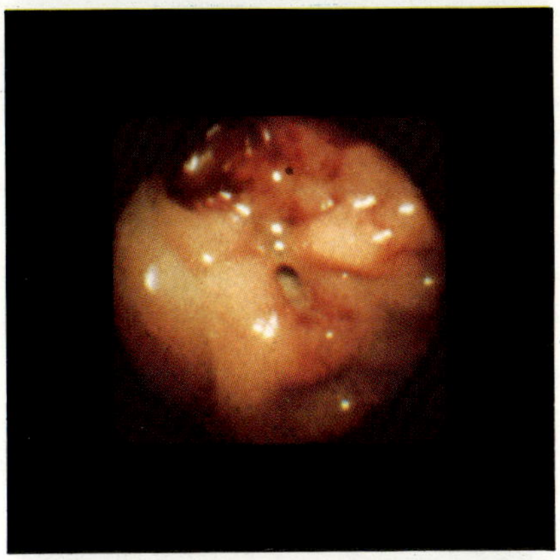

Fig. 132.
Sigmoiditis and diverticulosis. Small diverticulum on a hyperaemic sigmoid mucosa which forms high, oedematous, brittle, easily bleeding folds.

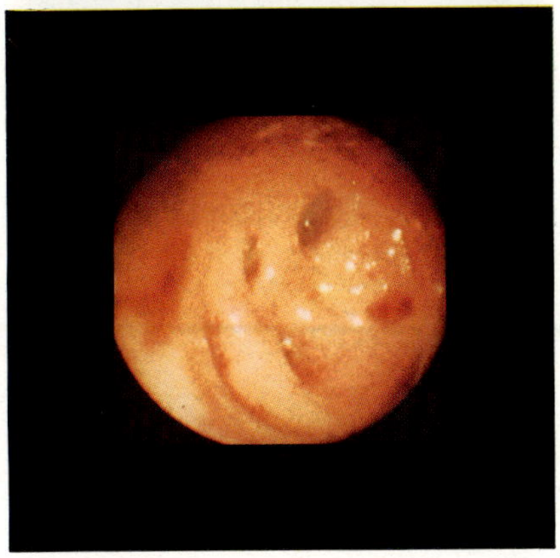

Fig. 133.
Diverticulitis. Small diverticula lined with very hyperaemic mucosa containing bloody serous material stand out from the congested mucosa of the sigmoid colon.

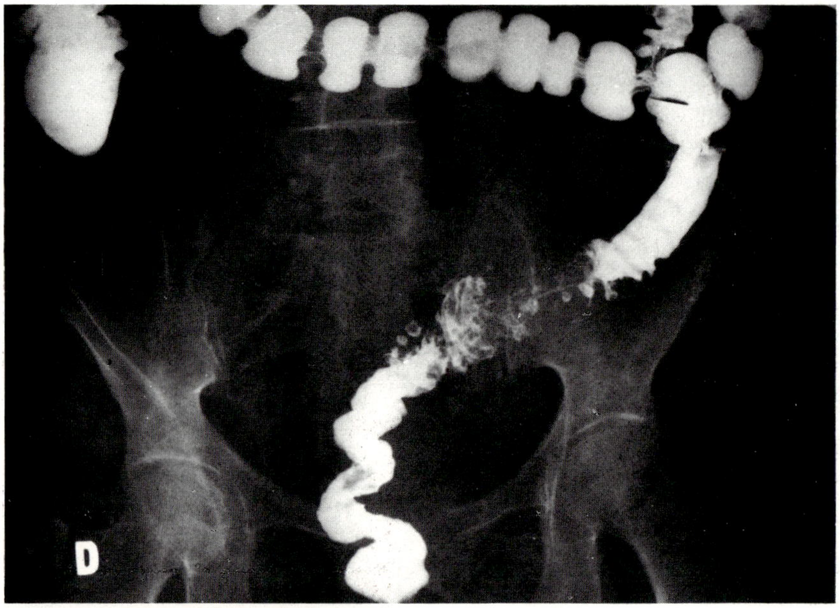

Fig. 134.
Barium enema. Diverticulitis. The descending colon is shortened, without haustrations and with irregular contours. The sigmoid-descending colon junction is narrowed, with a disorderly network of irregular folds. Numerous small, unevenly opaque diverticula can be seen in the deformed haustral pouches.

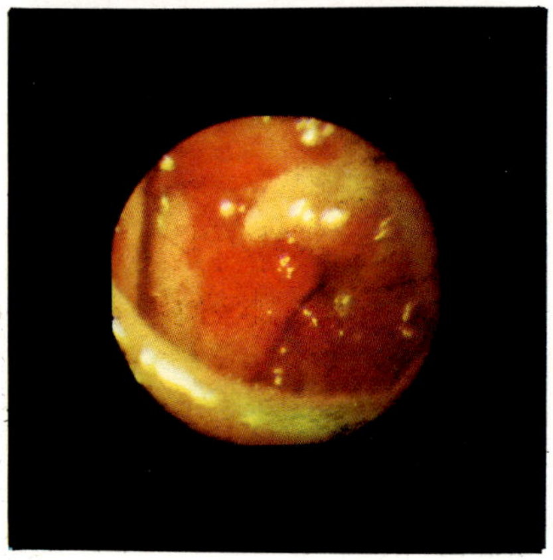

Fig. 135.
Same case. Diverticulitis and neoplastiform sigmoiditis. Large, bright red, apparently infiltrated, oedematous folds can be observed at the sigmoid-descending colon junction. The folds are elastic on palpation with the biopsy forceps. The bowel lumen is clearly narrowed.

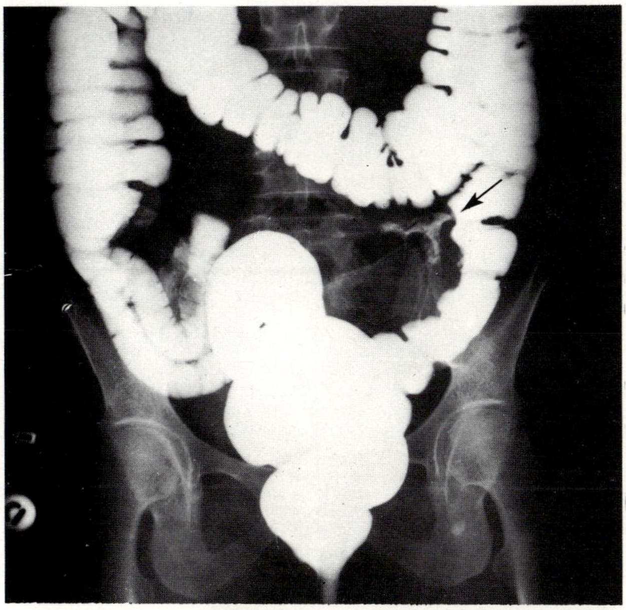

Fig. 136.
Barium enema. Fistula between the bowel and peritoneal cavity. The orifice of the fistulous passage is on the medial wall of the middle portion of the descending colon. The contrast medium passes into the peritoneal cavity.

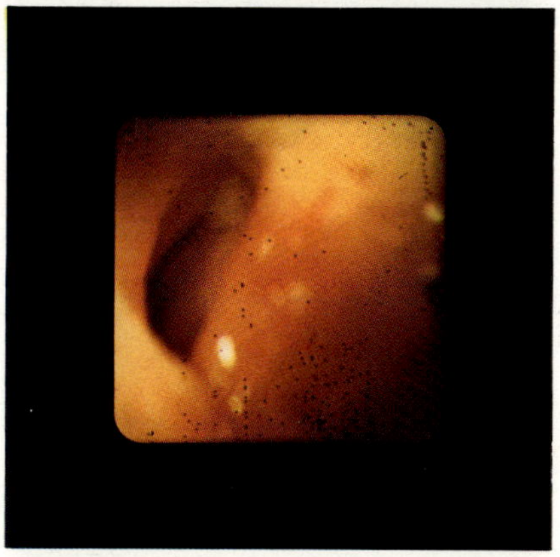

Fig. 137.

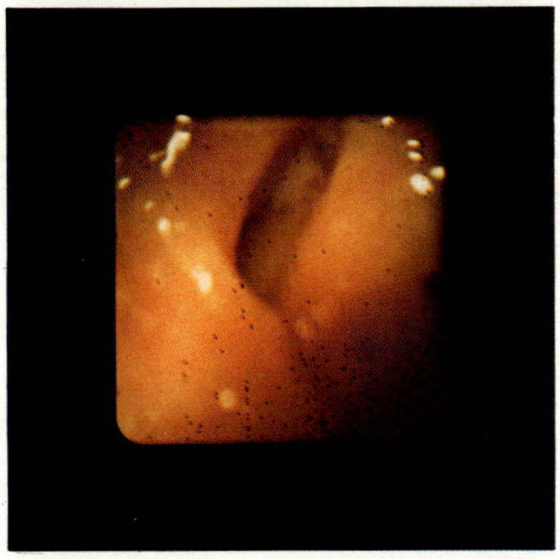

Fig. 138.

Figs. 137 & 138.
Fistulized diverticulum.
A large diverticulum with an oval orifice, lined with normal mucosa, can be seen in the middle third of the descending colon. The origin of the fistula was demonstrated fluoroscopically by injecting the contrast medium into the sacciform diverticular cavity.

Pneumatosis cystoides intestinalis

Fig. 139.
Direct abdominal X-ray. The descending colon is excessively distended by gas and shows numerous vacuolar pictures of bullous aspect.

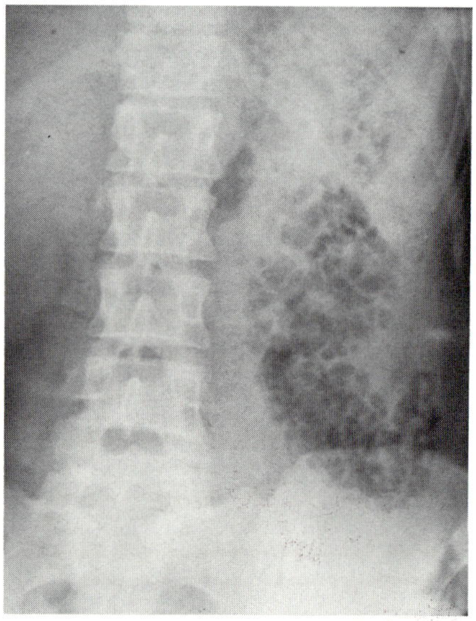

Fig. 140.
Soft elevations covered with smooth, shiny, pinkish mucosa can be seen on the bowel wall. The lesions are single or grouped and have the appearance of sessile polypoid growths. The adjacent mucosa is not infiltrated.

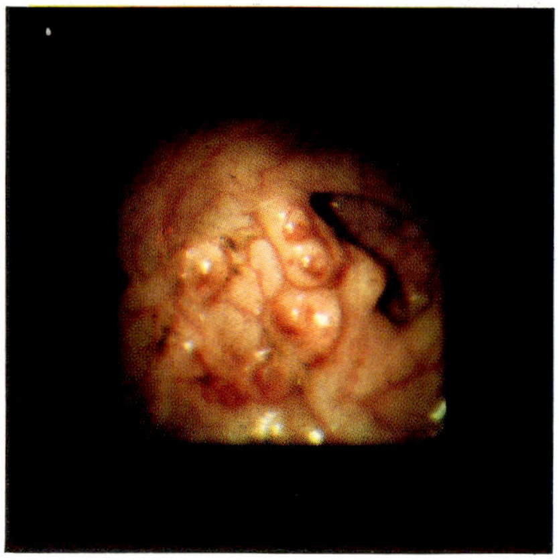

g. 141.
ndoscopic inspection shows that
e lesions are broad-based vesicles
nging in size from 0.2 to 0.8 cm.
ne impression is that some of these
sicles are intercommunicating and
at their content is easily displaced
contact with the fibrescope or on
lpation with the biopsy forceps.

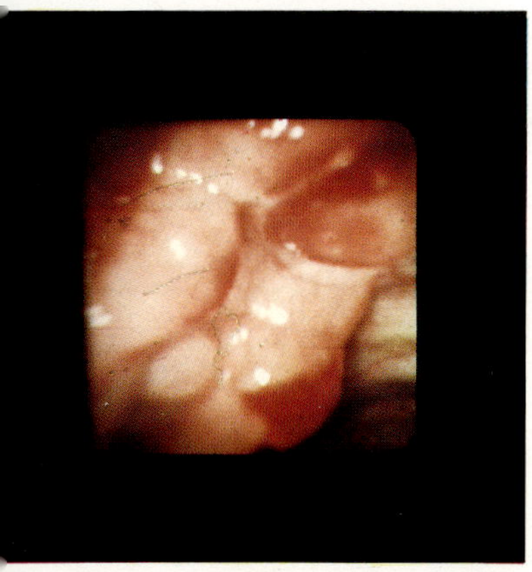

Fig. 142.
Same case. Biopsy on a cyst in the descending colon.

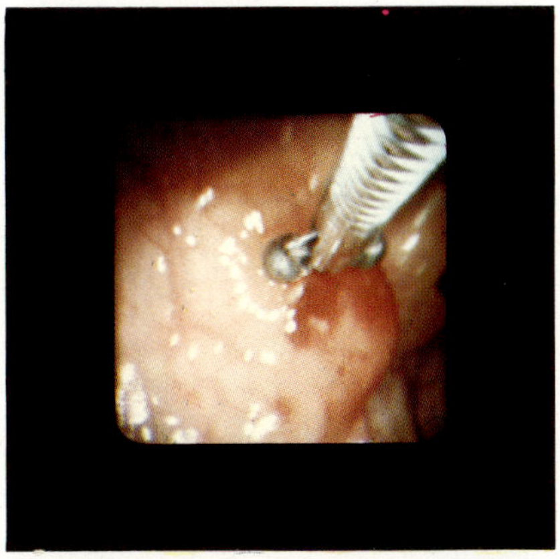

Fig. 143.
After biopsy the cyst collapses with a characteristic sound. The bowel wall containing the numerous cysts flattens after biopsy.

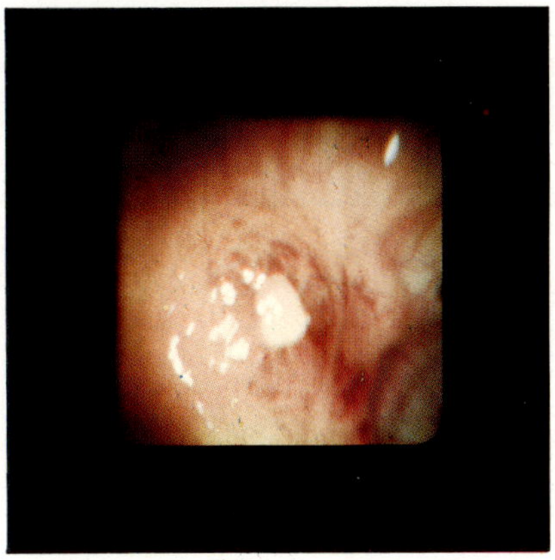

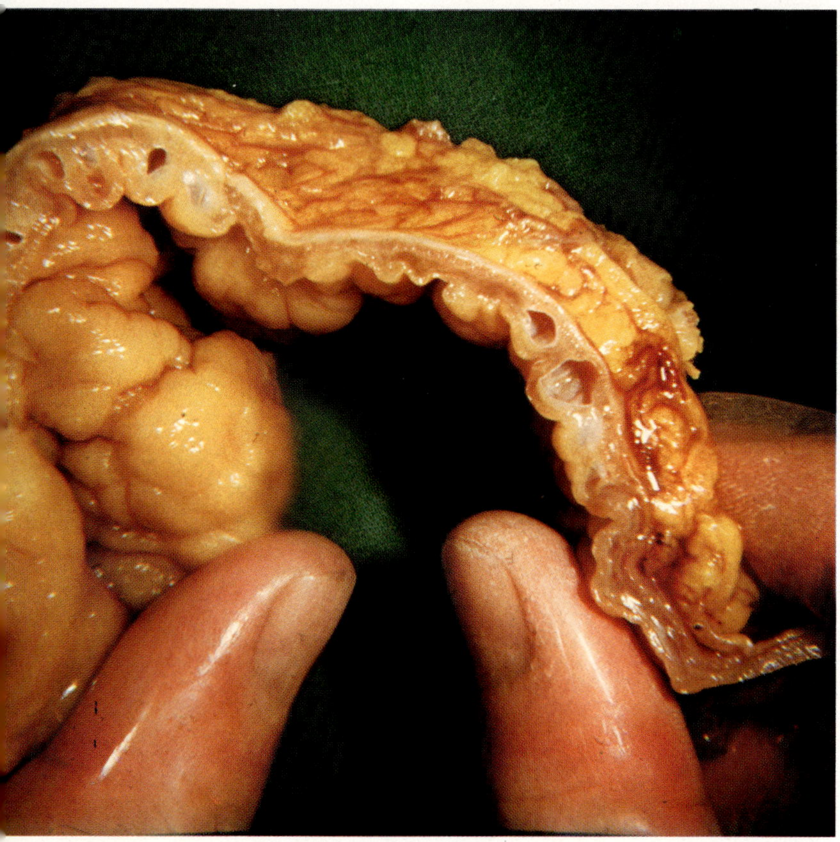

Fig. 144.
Surgical specimen. Numerous vacuolar lesions in the bowel wall. Pneumatosis cystoides intestinalis of the submucous variety.

(Figs. 139, 140, 141, 142, 143 & 144 by courtesy of dr. J. R. Armengol-Mirò, Barcelona).

Amoebiasis

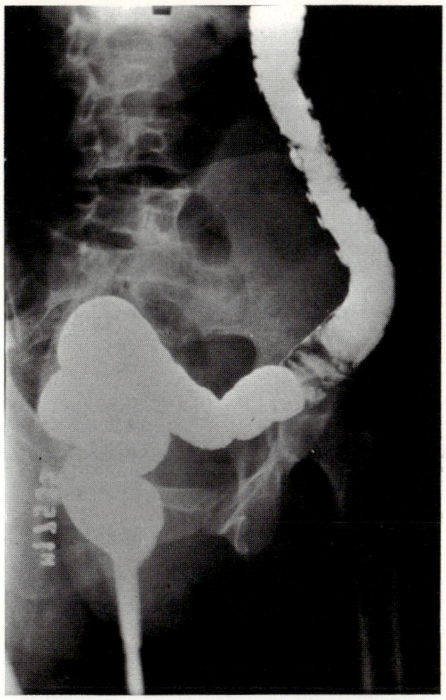

Fig. 145.
Barium enema. The descending colon is narrowed and no haustrations are apparent. The bowel wall is stiff and has irregular, serrated contours, some of the serrations being flat and collar stud-shaped.

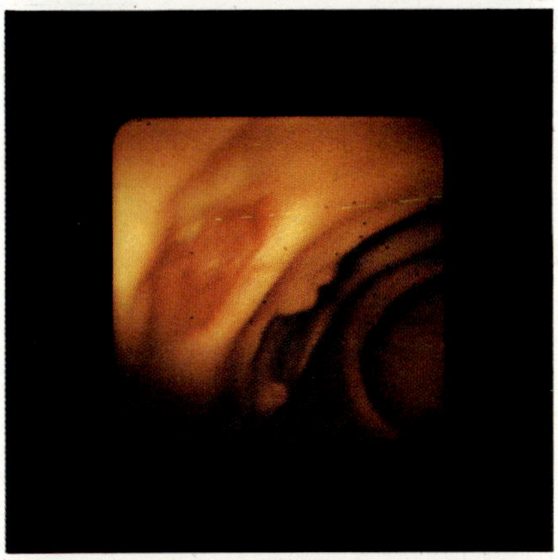

Fig. 146.
Round ulcers with undermined borders, about 1 cm in diameter, covered with whitish exudate can be seen through the coloscope in the descending colon. The mucosa between the ulcers is smooth, velvety and pink-coloured. Small pseudopolyps are also present.

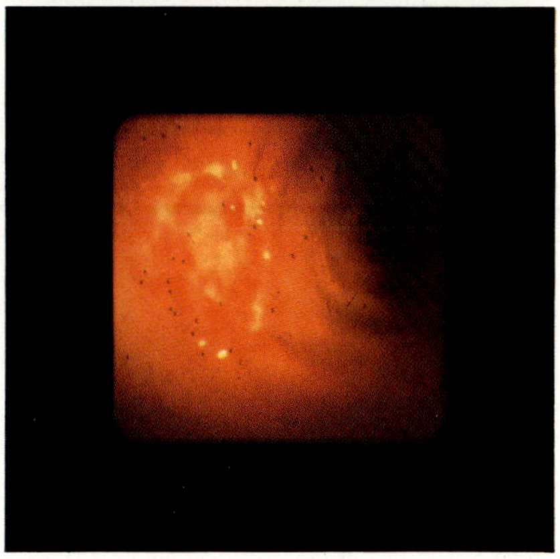

Fig. 147.
Same case. Detail of the granulations on the bottom of the ulcer.

The amoebic nature of the ulcers was ascertained only after Entamoeba histolytica had been fonud in samples of faeces taken during coloscopy.

Oxiuriasis

inworms (Oxyuris) are not uncommonly observed on the owel wall during coloscopy, from the sigmoid colon to the ecum. They may be single or grouped together and can asily be recognised by their small size (length 0,5-1 cm), reyish-white colour and active movements.

ig. 148.
inworm in the sigmoid colon.

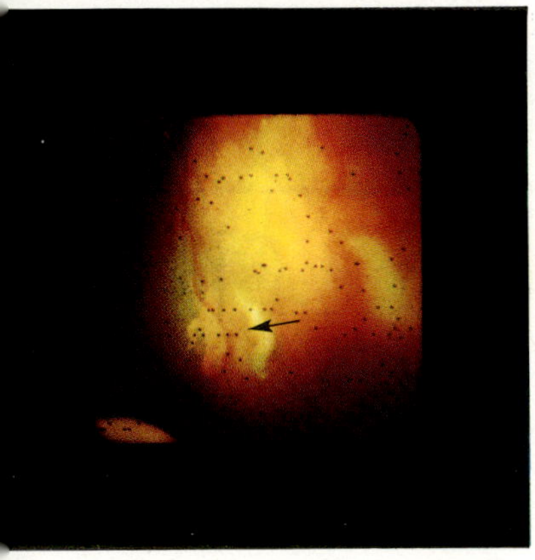

Fig. 149.
Two pinworms in the descending colon.

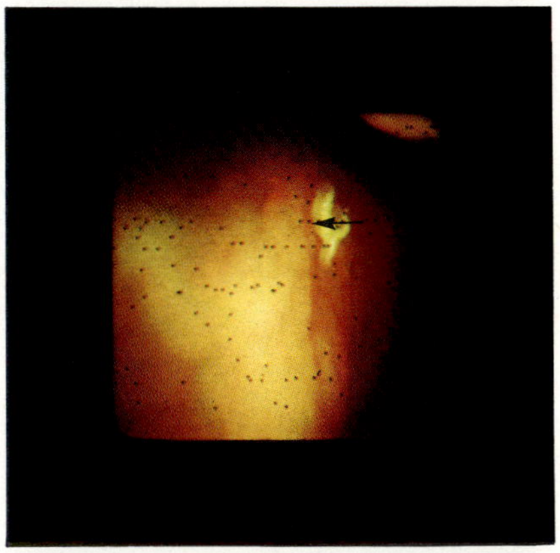

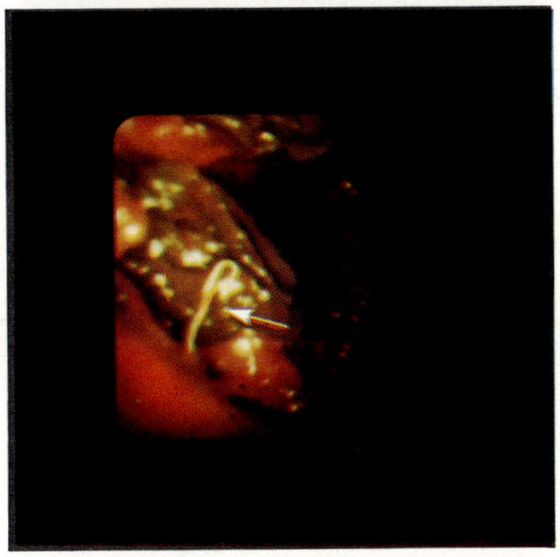

Fig. 150.
Pinworm in the cecum.

Tuberculosis of the colon

Fig. 151.
Barium enema.
The length of the right section of the colon is reduced. There is severe rigid narrowing of the cecum and ascending colon associated with irregular prestenotic dilatation of the last loop of the ileum which appears to be infiltrated. Granuloma-like filling defects are visible in the shrunken cecum. Two ulcers can be seen, one on the outer contour of the cecum and another on the upper contour of the prevalvular portion of the ileal loop.

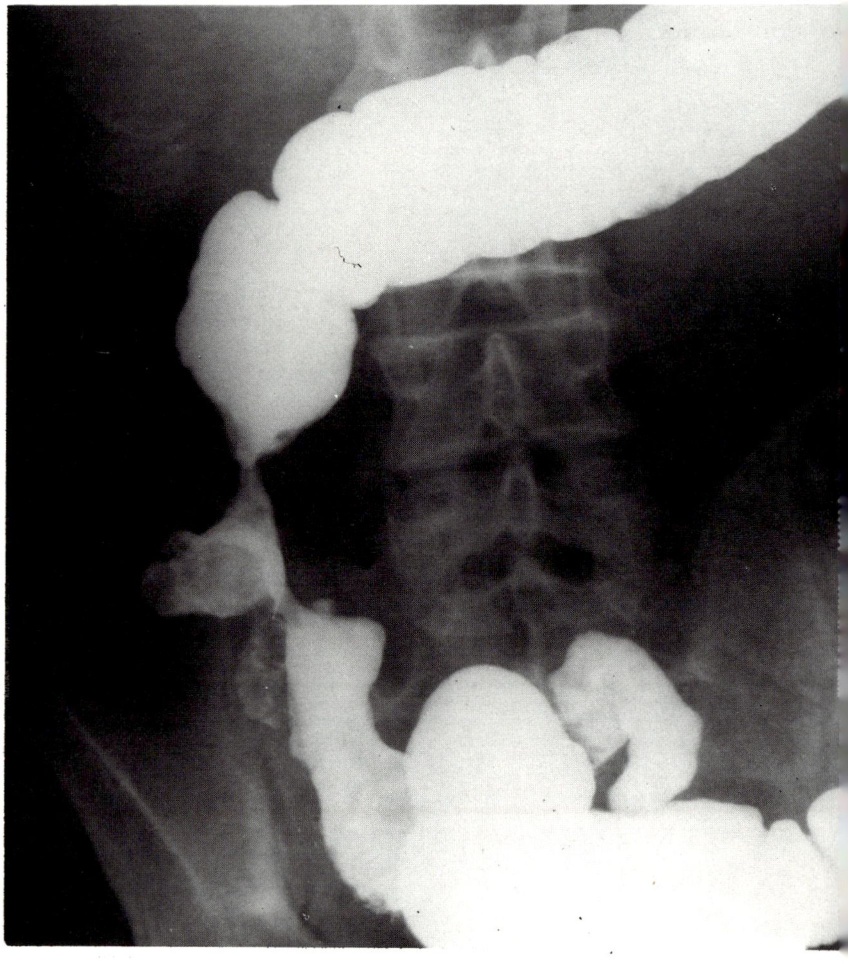

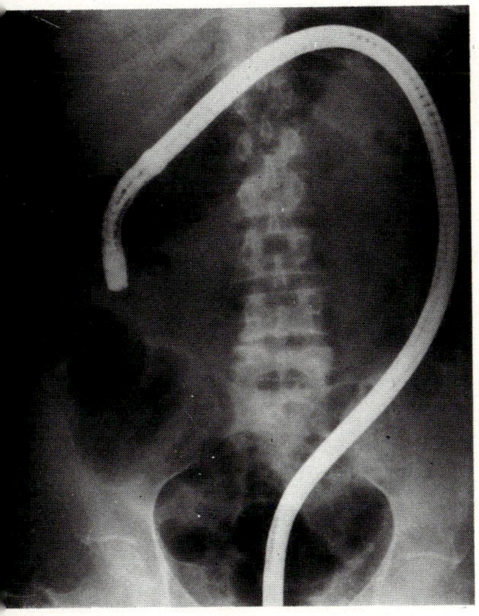

ig. 152.
he coloscope has been arrested in
ie ascending colon at the stenosis
:en on barium enema.

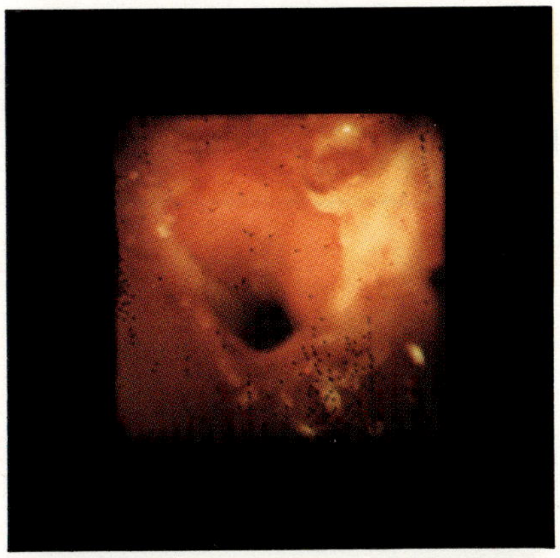

Fig. 153.
Partial stenosis of the ascending colon. The bowel wall is stiff and infiltrated. Large ulcerated patches with irregular, undermined margins, covered with whitish exudate, are scattered over the congested mucosa.

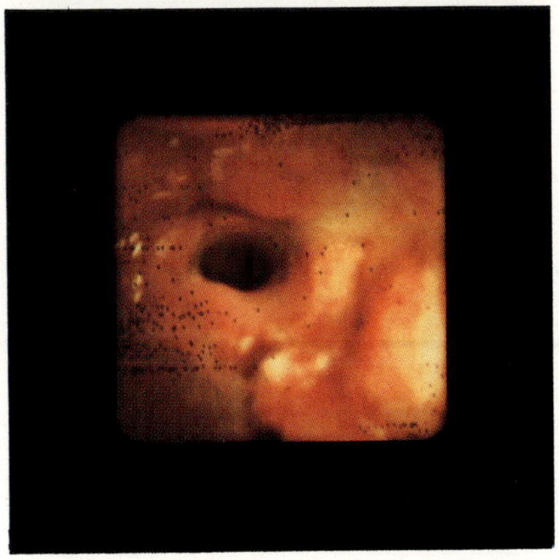

Fig. 154.
Detail of the same case showing smooth, shiny bright red, elevations between the ulcers. Histologic preparations of material taken with the biopsy forceps from one of the prestenotic elevations confirmed the diagnosis of tuberculosis of the large bowel.

Fig. 155.
Partial stenosis of the transverse colon. The bowel wall is stiff and infiltrated. The mucosa is greyish-pink in colour. There are two large ulcers with irregular margins. Bleeding after biopsy was slight. The histologic diagnosis was tuberculosis of the large intestine. (Courtesy of prof. A. Montori, Rome.)

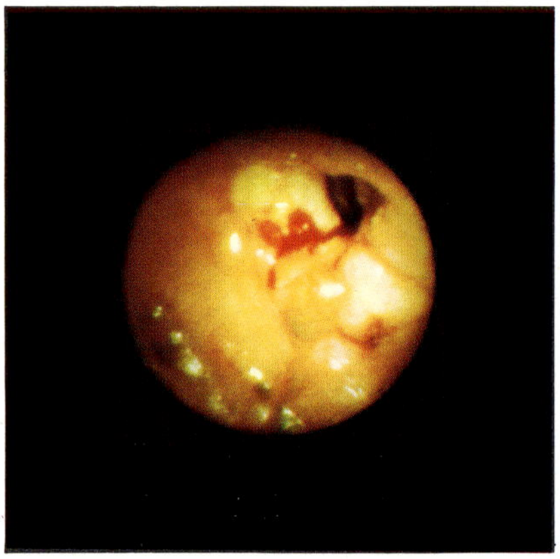

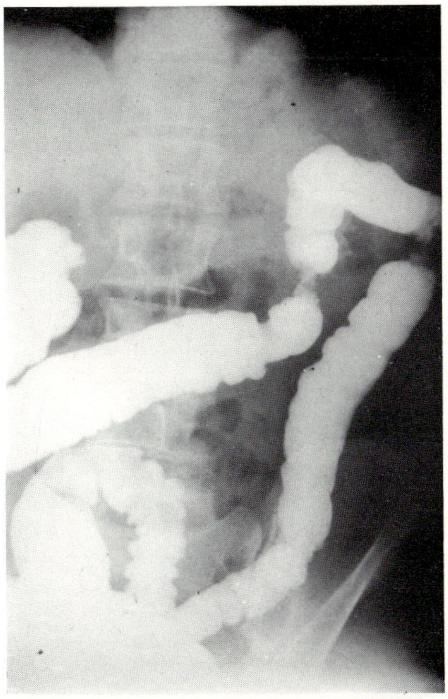

Fig. 156.
Barium enema. Stricturing of the large intestine can be seen in the transverse colon, close to the splenic flexure, and at the beginning of the descending colon. (Courtesy of dr. Armengol-Miro, Barcelona.)

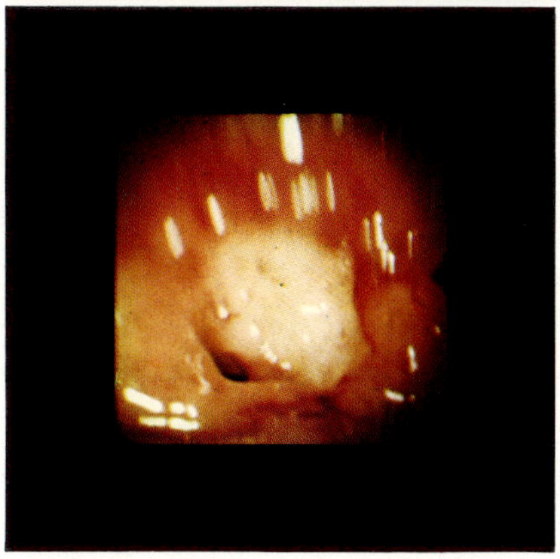

Fig. 157.
Substenosis of the splenic flexure. Endoscopic view. A hard, greyish patch with irregular margins can be seen at the splenic flexure. The surrounding mucosa is infiltrated. The bowel wall is stiff and the bowel lumen is narrowed, making further progression of the fibrescope impossible. The histologic diagnosis was tuberculosis of the large intestine. (Courtesy of dr. Armengol-Miro, Barcelona.)

Idiopathic ulcerative rectocolitis

:cute idiopathic ulcerative rectocolitis.

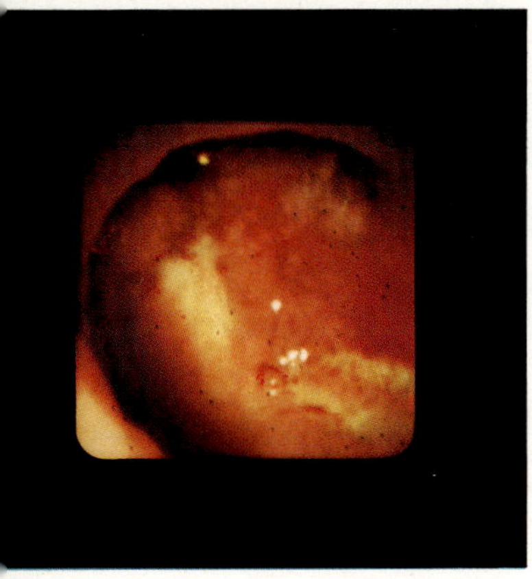

;. 158.

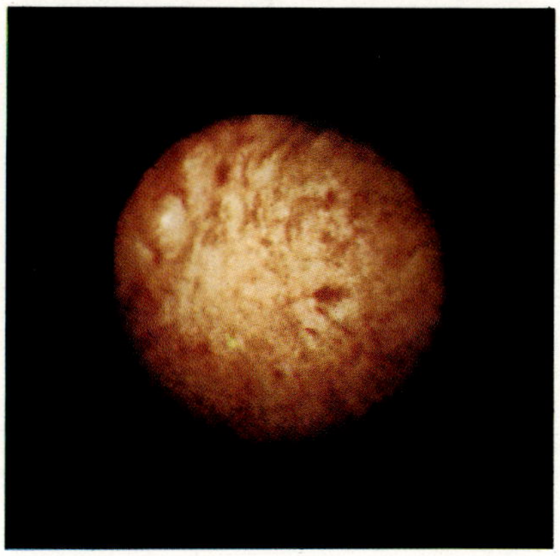

Fig. 159.

The mucosa is diffusely oedematous, intensely congested, shiny, brittle to the touch and bleeds easily.

Fig. 158.
Capillary haemorrhages.

Fig. 159.
Blood oozing from numerous shallow erosions on the mucosa.
The endoscopic picture is that of an inflammatory condition in the active stage.

.active chronic idiopathic ulcerative rectocolitis.

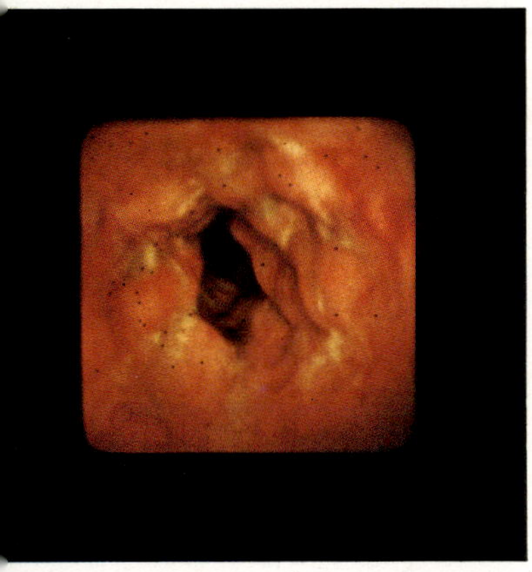

;. 160.

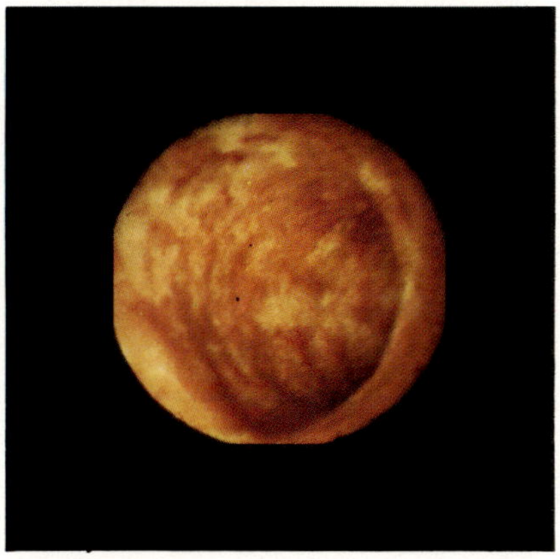

Fig. 161.

Fig. 160.
The mucosa of the descending colon is diffusely hyperaemic and moderately oedematous. Various anatomical changes are apparent. The normal configuration of the colonic haustrations is altered and there are pseudopolyps and depressions due to scarring where ulcers were formerly present.
Fig. 161.
The mucosa of the descending colon is hyperaemic, granular and dry "like sandpaper", and is brittle to the touch.
The endoscopic picture is that of inactive chronic idiopathic ulcerative rectocolitis.

:ctive chronic idiopathic ulcerative rectocolitis.

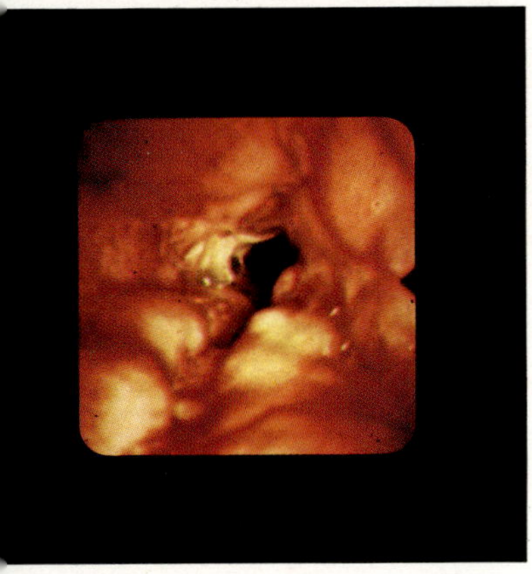

;. 162.
ie wall of the descending colon is inted with bright red blood. The ustral folds have disappeared. iere are numerous brittle eudopolyps.

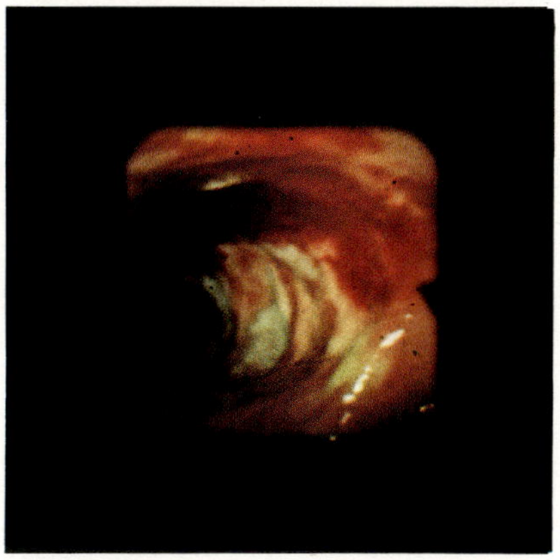

Fig. 163.
The mucosa is covered with a layer of mucus and fibrin. The entire mucosa which lines the bowel wall oozes blood.

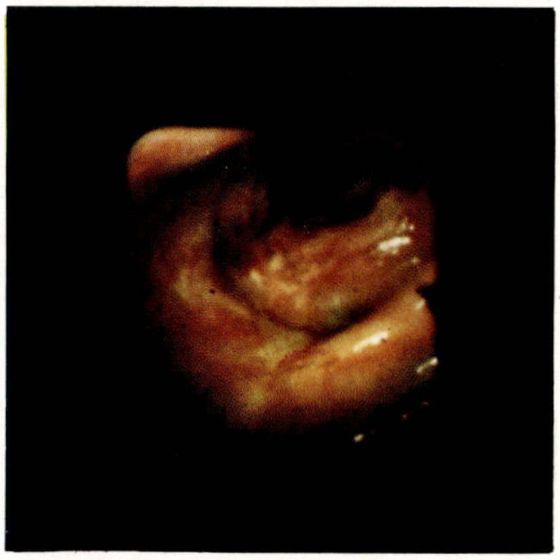

Fig. 164.
The mucosa is oedematous, hyperaemic, and pitted with numerous ulcers of various shapes.

Early, mildly active idiopathic ulcerative rectocolitis.

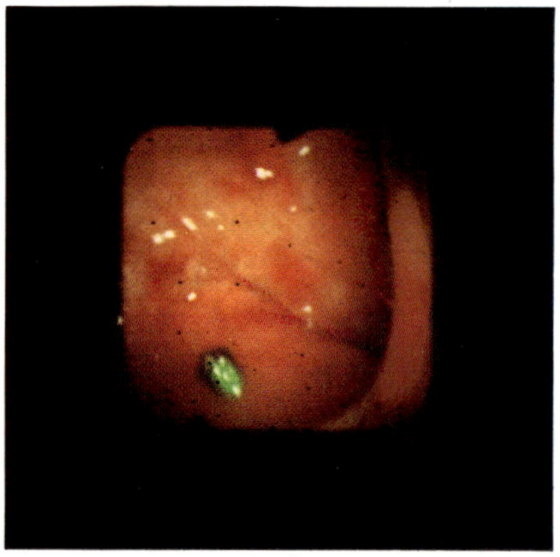

Fig. 165.
The underlying submucous vascular network shows up through the mucosa which is oedematous, congested, with areas of intense hyperaemia, moist and shiny.

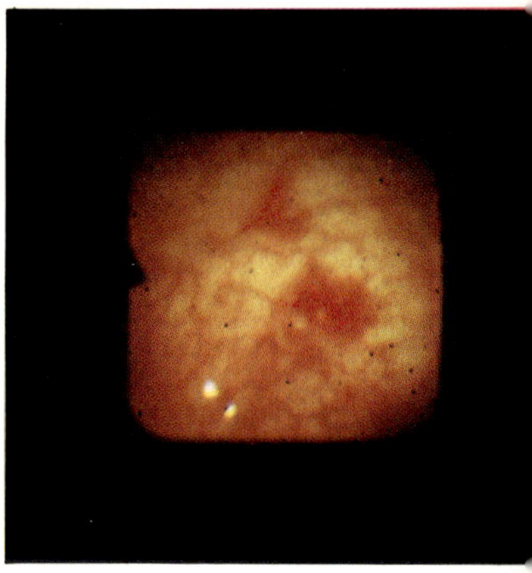

Fig. 166.
Closer view of the same case.

Fig. 170.
Barium enema. The bowel wall is stiff and no haustrations are visible. The contours of the bowel, especially those of the transverse colon, have a tattered or serrated aspect.

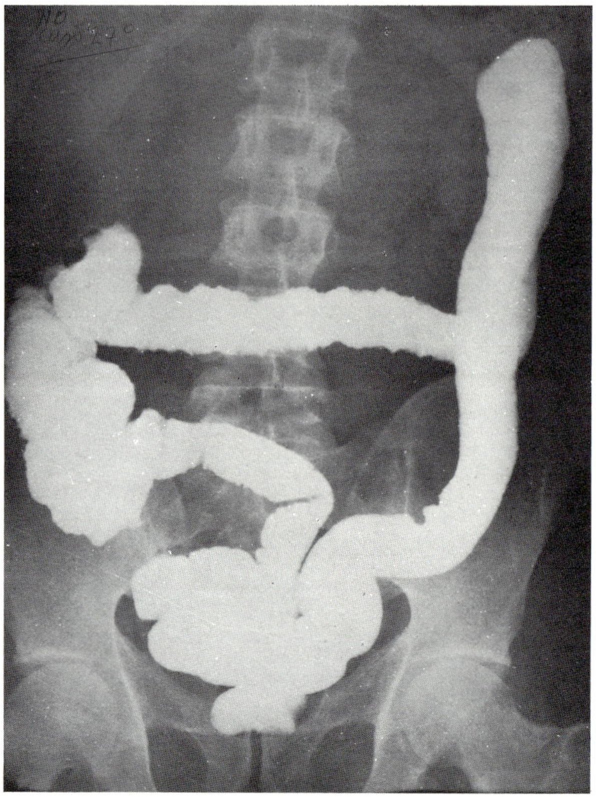

Figs. 171 & 172.
The mucosa of the sigmoid colon is hyperaemic. The submucous vascular network is clearly visible. Bleeding is caused by the fibroscope.

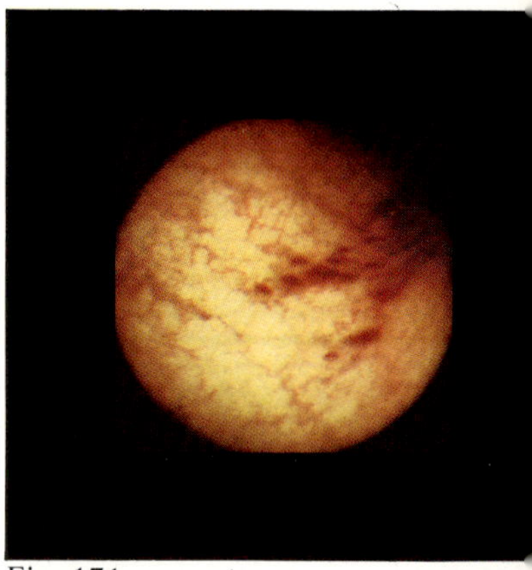

Fig. 171.

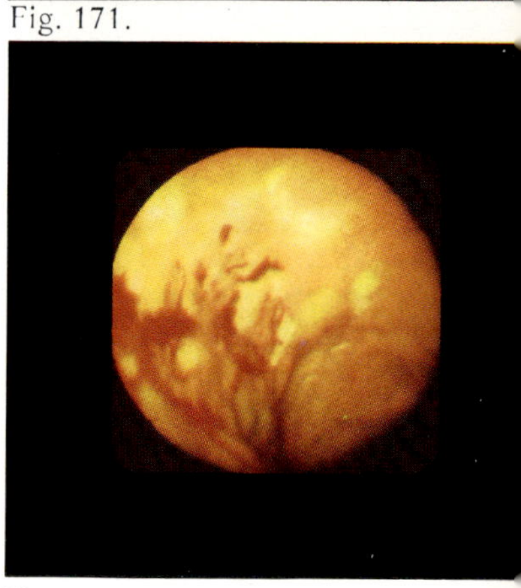

Fig. 172.

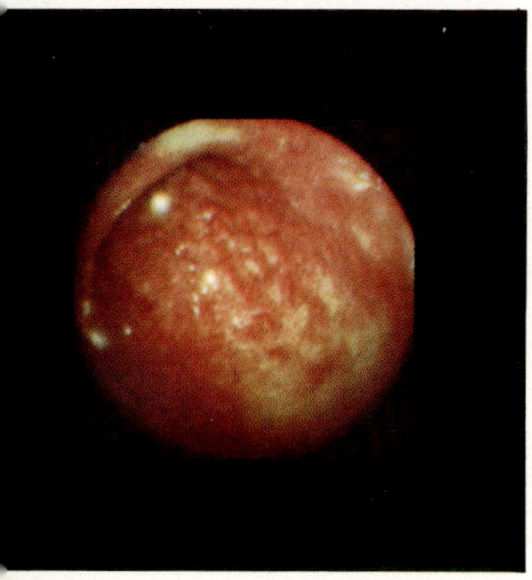

:. 167.
e mucosa of the descending
:tion of the sigmoid colon is
:raemic, granular and marred by
llow erosions.

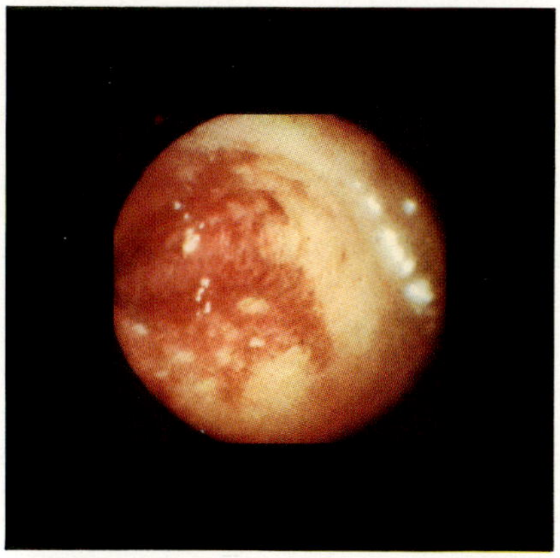

Figs. 168 & 169.
At the sigmoid-descending colon junction the inflammatory process becomes milder. Note the boundary line between the inflamed mucosa and the apparently normal mucosa above.

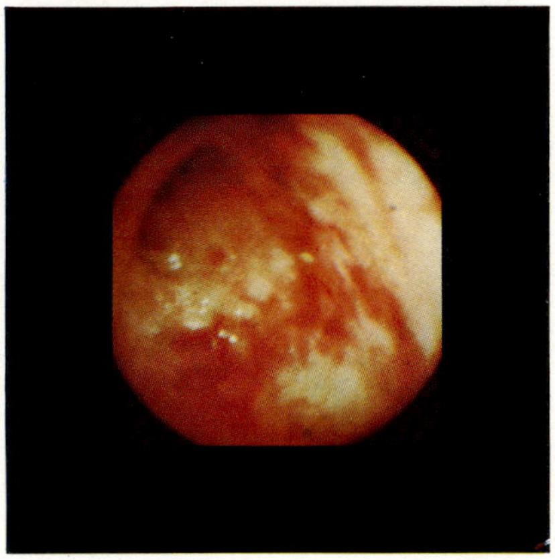

Fig. 169.

ʒ. 173.

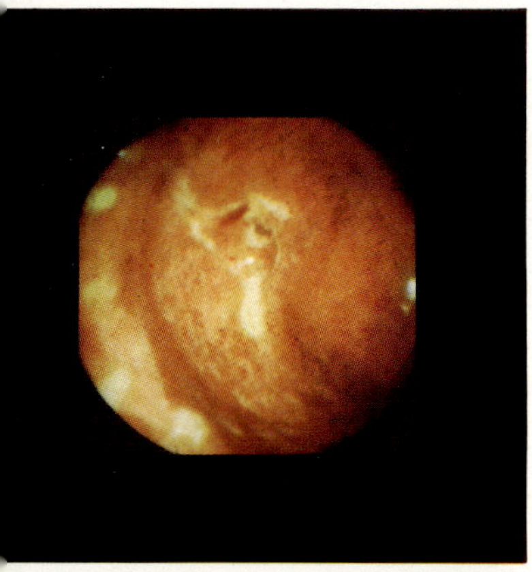

Fig. 174.

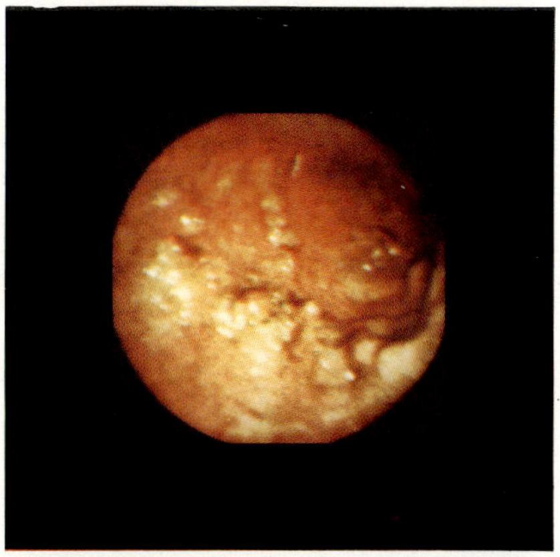

Fig. 173.
The mucosa of the distal portion of the descending colon is hyperaemic, thin and granular in aspect. The bowel wall is covered with a layer of mucus.

Fig. 174.
Middle third of the descending colon. The bowel wall is distensible and lined with hyperaemic mucosa. Numerous small, brittle pseudopolyps protrude into the bowel lumen. Collections of serofibrinous material are also visible.

Fig. 175.
Left flexure. The bowel wall is less distensible. Numerous pseudopolyps are present. There are no haustrations.

Fig. 176.
The wall of the transverse colon is distensible and lined with a finely granular mucosa which forms low folds (the haustral folds are preserved). There are numerous erosions and "en nappe" haemorrhages.

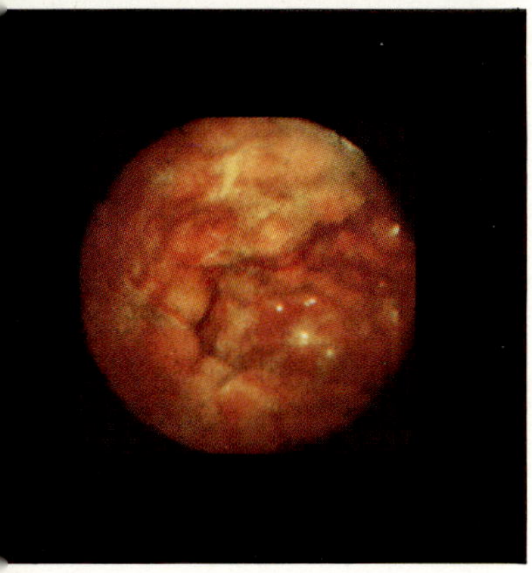

ʒ. 175.

Fig. 176.

ıe endoscopic picture suggests moderately active chronic iopathic ulcerative rectocolitis.

Fig. 177.
Barium-enema. The contours of the colon are spiky, especially at the hepatic flexure. Its surface has a marbled, granular aspect, with a mosaic-like pattern in the transverse colon due to the simultaneous presence of pseudopolyps and small ulcers.

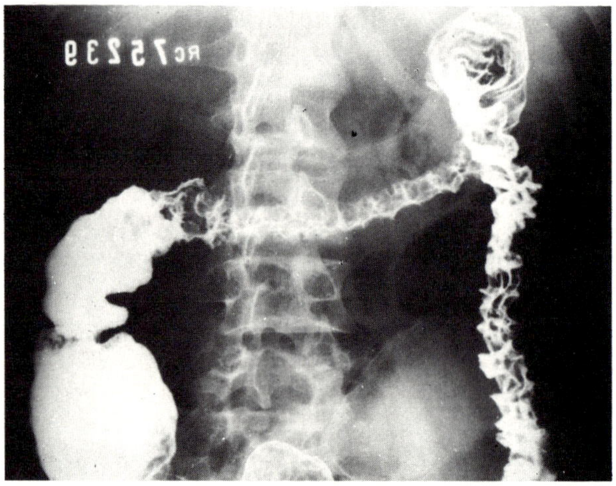

;. 178.
vere anatomical changes in the lon—the haustrations have appeared and there are gross eudopolyps together with various pes of ulcers. There is fresh blood the bowel wall.

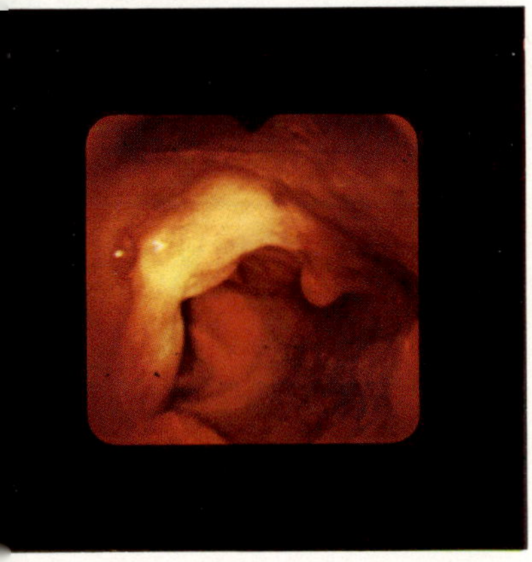

Fig. 179.
Same case. The wall of the transverse colon is not very distensible. There are pseudopolyps protruding into the bowel lumen.

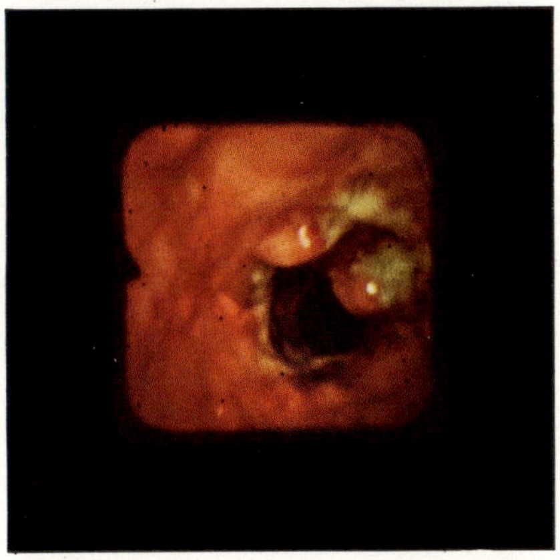

Fig. 180.
At the hepatic flexure the bowel lumen is obstructed by a mushroom-like growth which infiltrates the bowel wall and impedes further progression of the fibrescope. A coloscopic biopsy on the growth showed it to be an adenocarcinoma.

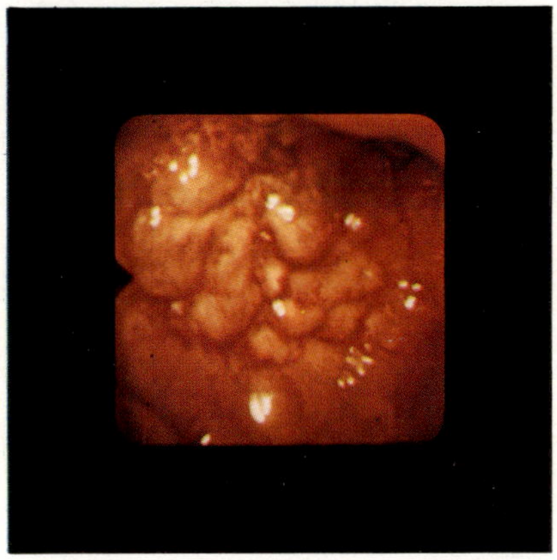

he coloscopic picture is that of active chronic idiopathic lcerative rectocolitis associated with adenocarcinoma of the epatic flexure.

Fig. 181.
Barium enema. The colon is uniformly narrowed and tubular, with small filling defects due to a rug-like array of pseudopolyps associated with microulcers of the mucosa, corresponding to small spikes on the bowel contour.

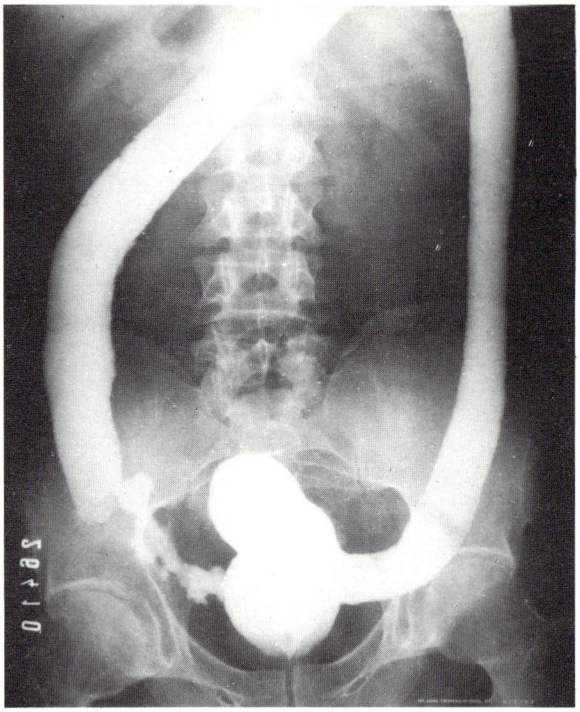

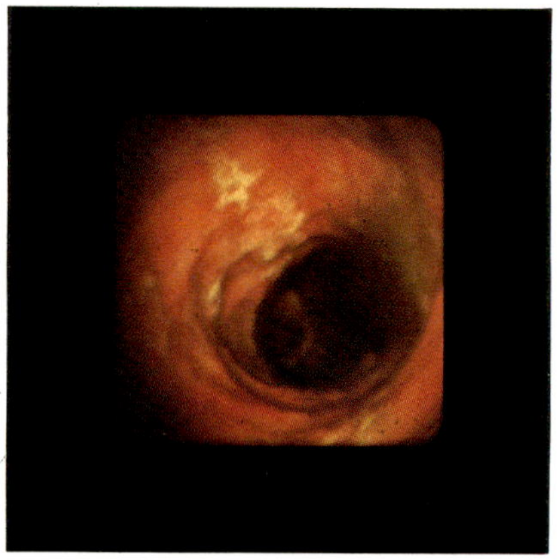

Fig. 182.

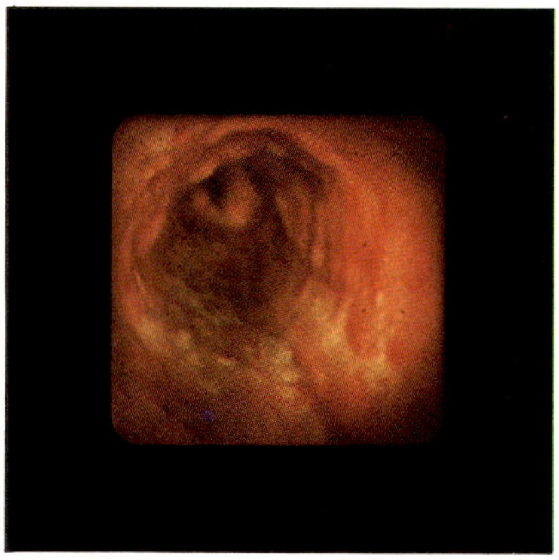

Fig. 183.

Figs. 182 & 183.
The descending colon is lined with a pinkish-red, granular mucosa with pseudopolyps. In this chronic stage of the disease the descending colon has a characteristic tubular or stove pipe-like aspect.

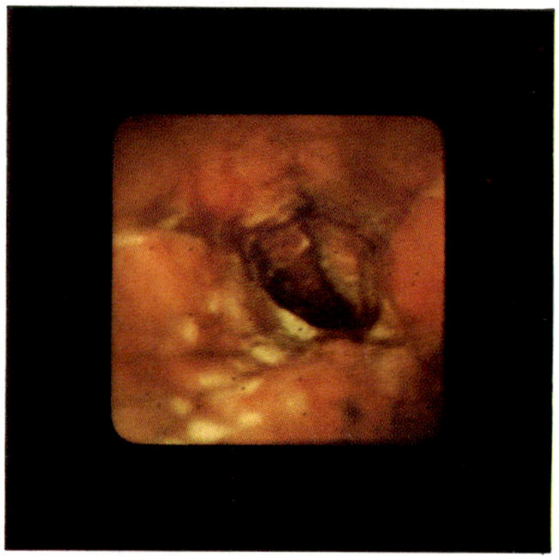

Fig. 184.
Numerous greyish depressions corresponding to crypt abscesses are strewn along the wall of the descending colon.

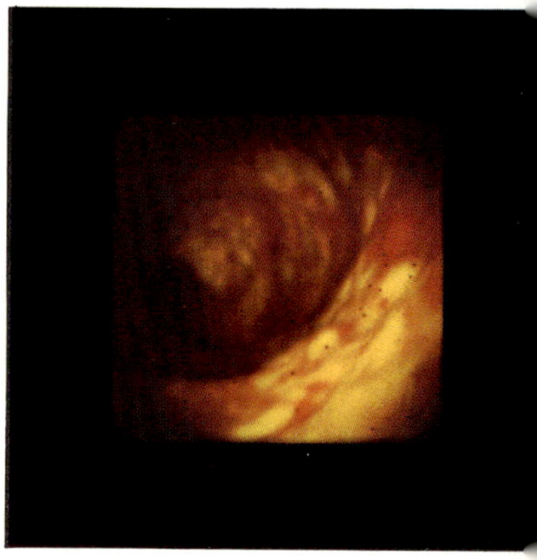

Fig. 185.
Closer view of the greyish depressions (crypt abscesses).

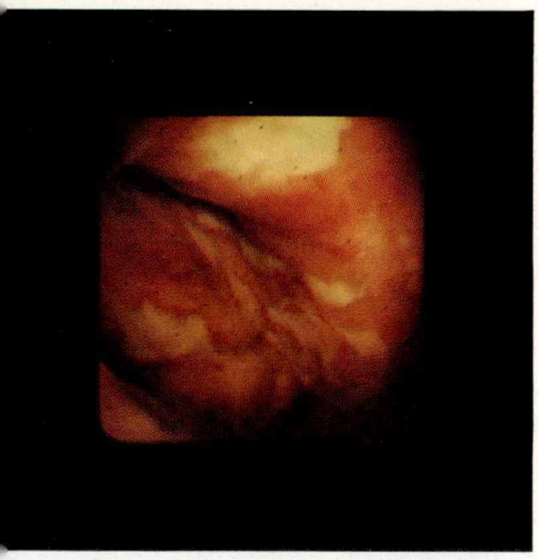

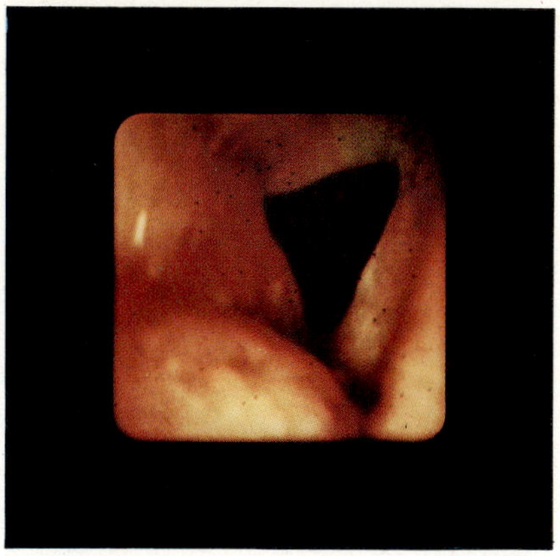

Fig. 186.
At the splenic flexure the mucosa is thin, smooth and shiny and is pitted with punched-out ulcers of various sizes.

Fig. 187.
Transverse colon. The triangular configuration of the haustral folds is barely recognizable.

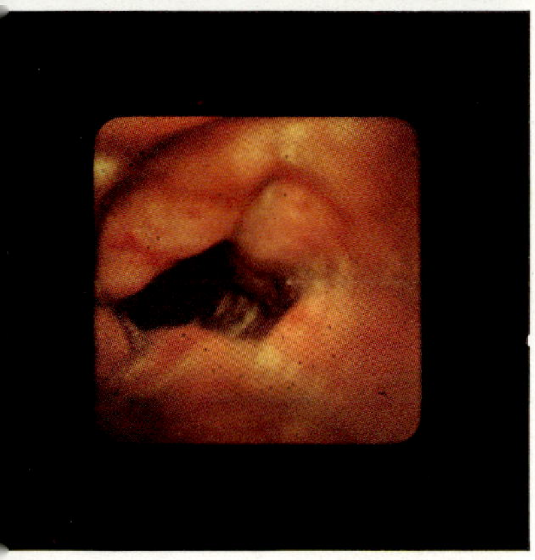

'ig. 188.
scending colon. The mucosal folds ave disappeared. The mucosa is nooth, thin and oedematous.

Fig. 189.
Fundus of the cecum. The mucosal folds have disappeared. The lumen is wide. Greyish depressions corresponding to crypt abscesses are visible on the pinkish-red mucosa.

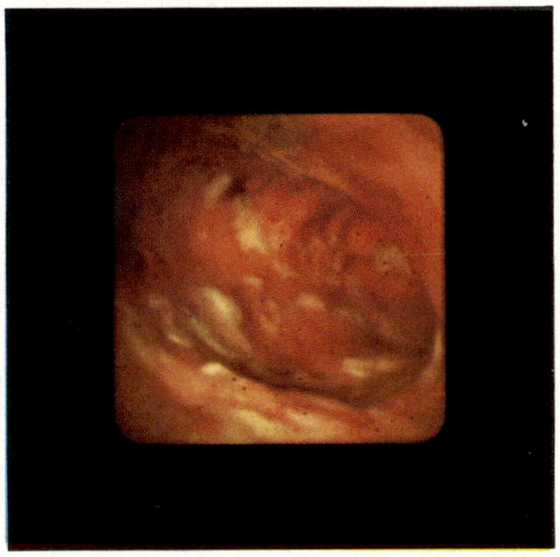

Fig 190.
Closer view of the gaping ileocecal papilla.

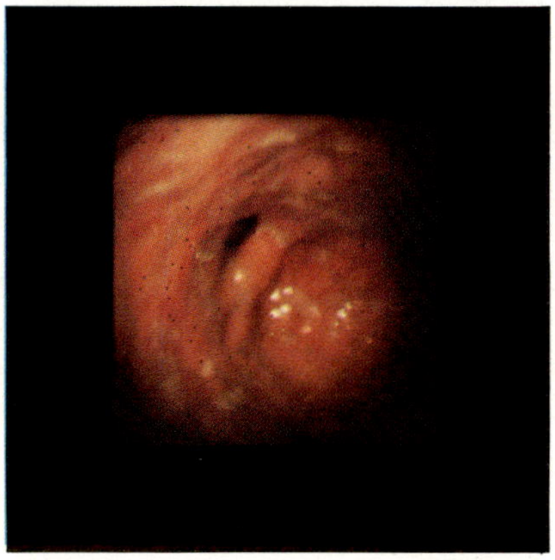

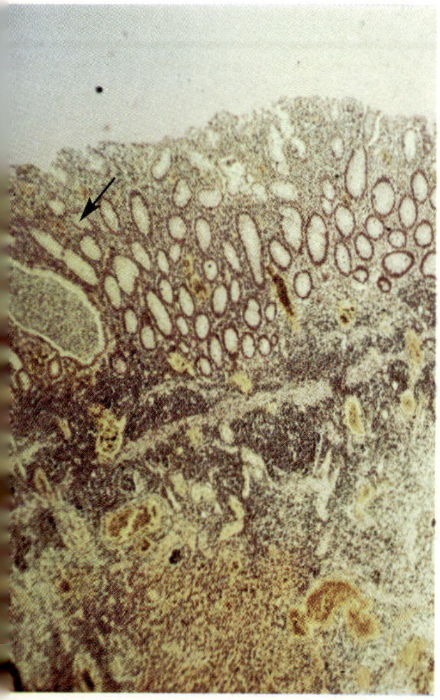

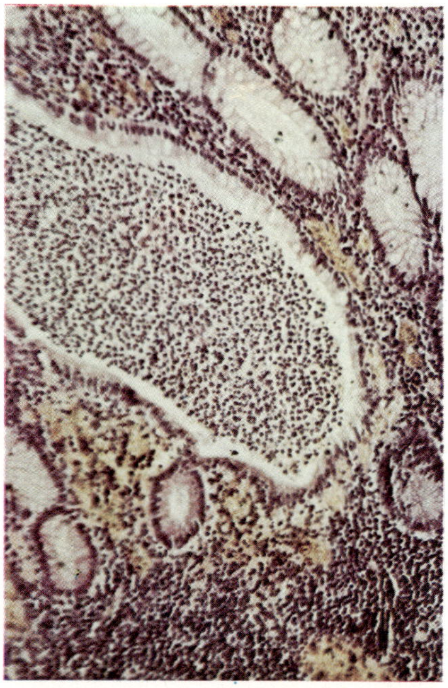

Fig. 191.
Coloscopic biopsy taken from one of the depressions of the mucosa of the descending colon. The histologic diagnosis was haemorrhagic ulcerative rectocolitis.

Fig. 192.
Detail of Fig. 191 with higher magnification. Crypt abscess.

The radiologic, coloscopic and histologic picture corresponds to that of dormient idiopathic ulcerative rectocolitis involving the whole colon.

Figs. 193, 194 & 195.
Endoscopic view of the ileocecal papilla in its three functional aspects: 1) open; 2) passage of faeces; 3) closed. The ileocecal papilla is affected by an inflammatory process. The mucosa is oedematous, shiny, moist and hyperaemic and contains numerous haemorrhages.

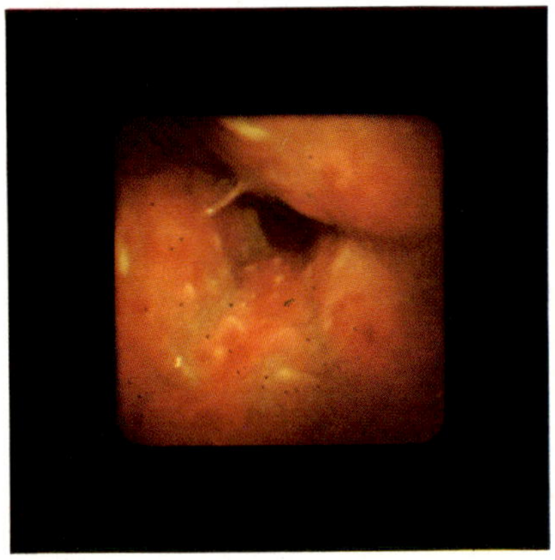

Fig. 194.

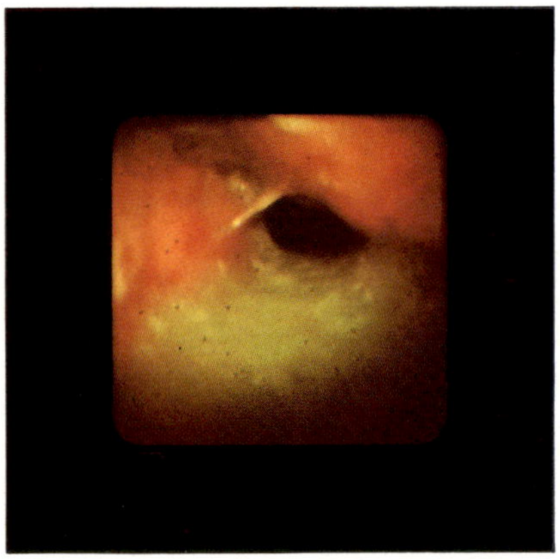

Fig. 195.

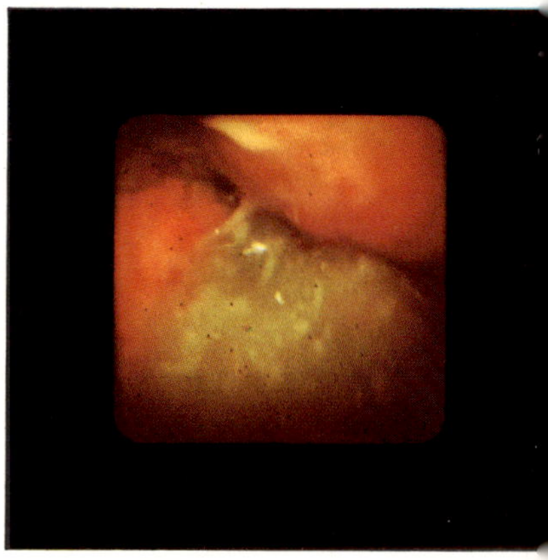

ig. 196.
diopathic ulcerative rectocolitis of
he ileocecal papilla. Numerous
inear ulcers are visible on the
ongested, oedematous mucosa.

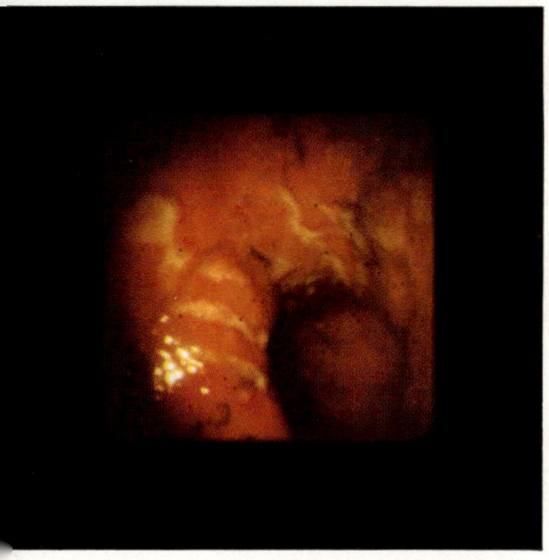

Fig. 197.
Closer view of the same case
showing intense oedema and
ulceration of the mucosa of the
papilla.

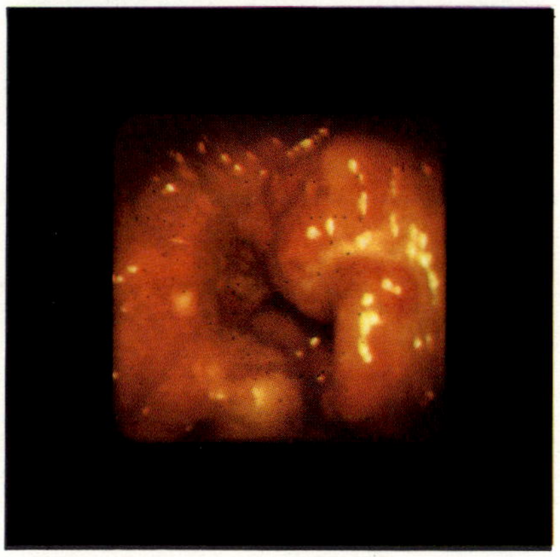

Fig. 198.
Same case. The inflammatory
process extends 6-8 cm into the
terminal ileal loop.

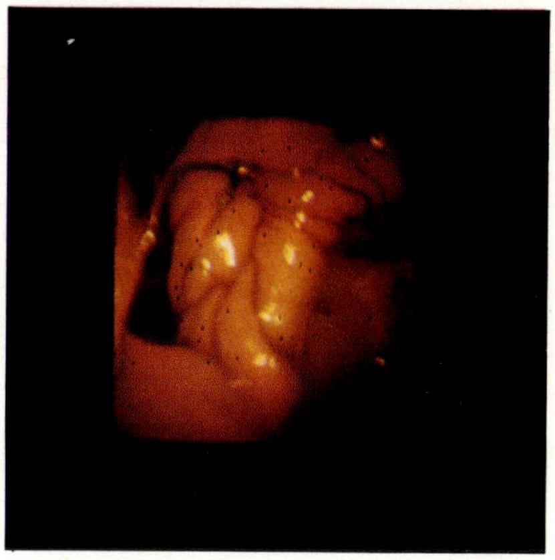

Pseudopolyps in idiopathic ulcerative rectocolitis.

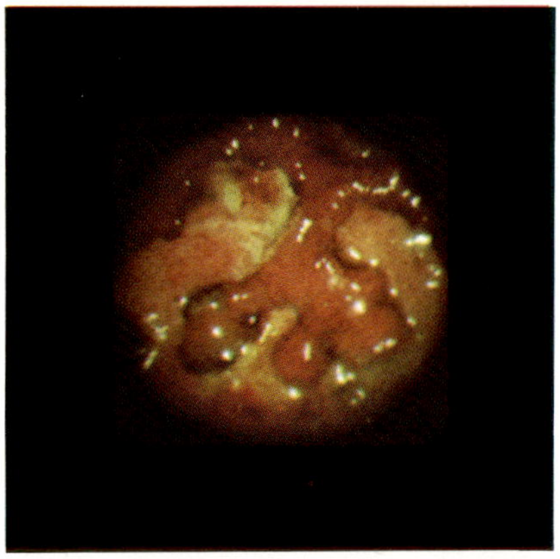

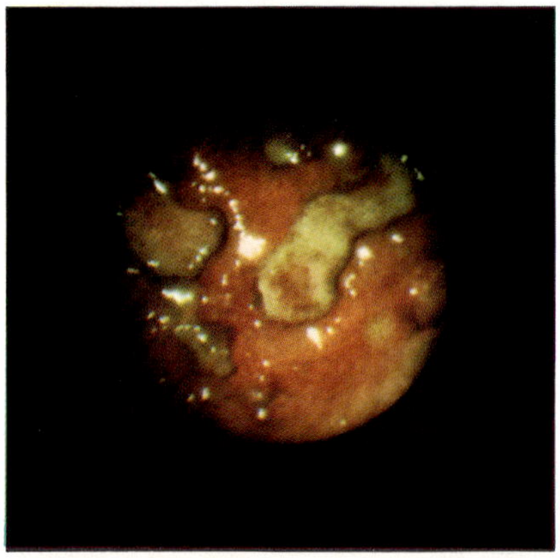

Figs. 199 & 200.
Pseudopolyps of monstruous aspect protruding from a hyperaemic, shiny mucosa containing many shallow erosions and covered in part by a layer of mucus and pus.

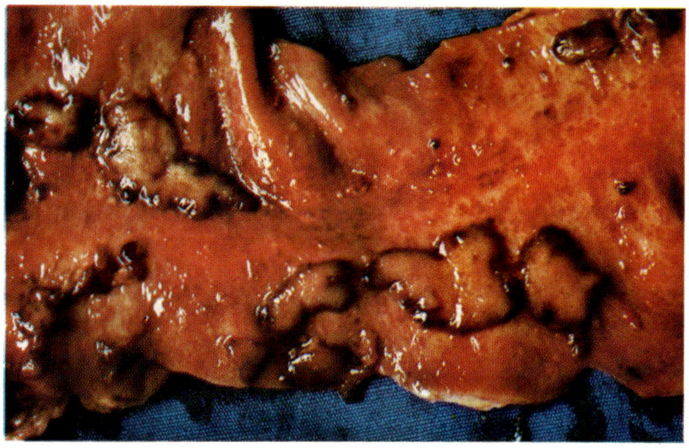

Fig. 201.
Surgical specimen from the same case. A length of colon of normal thickness. The haustral folds are lost. The mucosa contains a great number of erosions. Scattered patches of regenerating tissue.

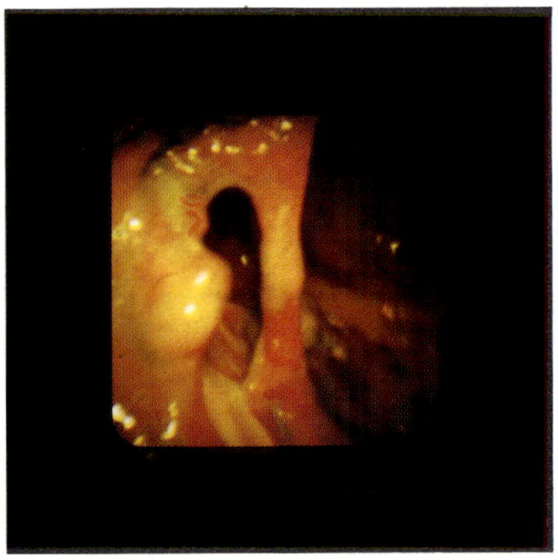

Fig. 202.
Pseudopolyps bridging the bowel lumen. (Courtesy of dr. J. R. Armengol-Miro, Barcelona.)

Figs. 203 & 204.
Numerous round-shaped growths, ranging in size from that of a millet seed to a pea, stand out from the mucosa of the sigmoid colon. The pseudopolyps are brittle when touched by the biopsy forceps. Blood oozes from the hyperaemic, shiny, brittle mucosa.

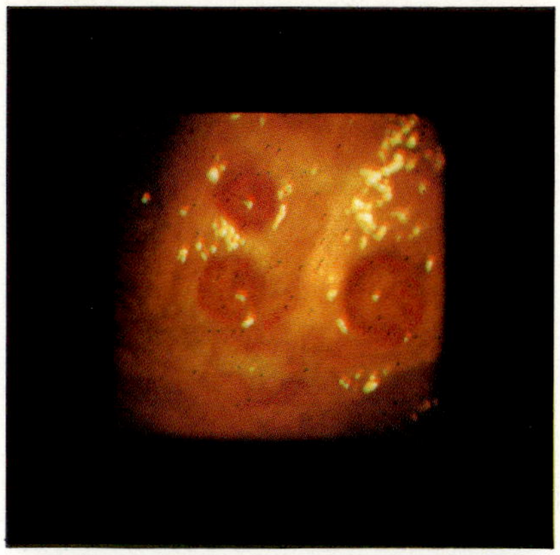

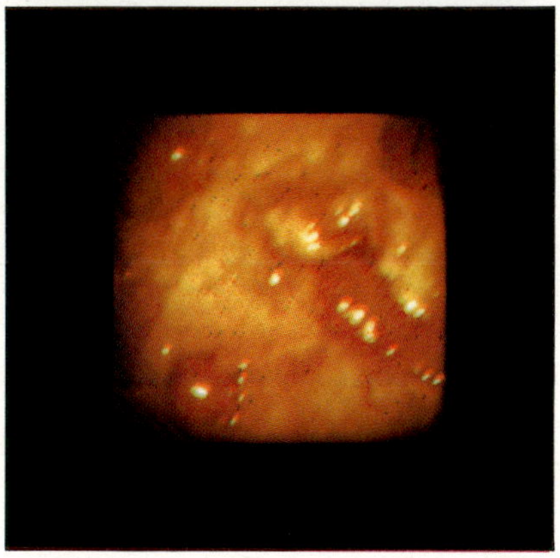

Fig. 204.

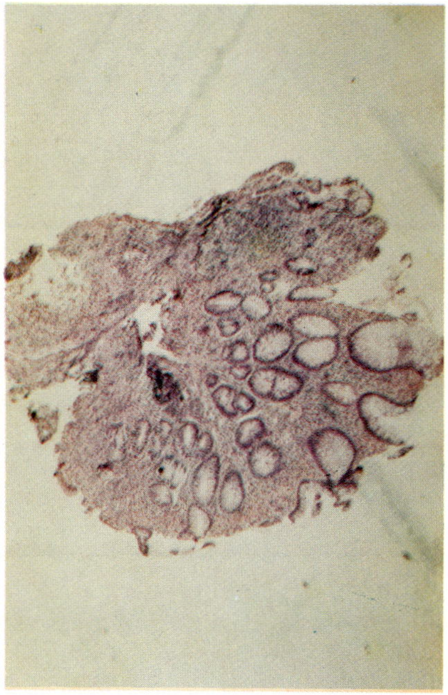

Fig. 205.
Coloscopic biopsy taken from a pseudopolyp in the same case. Histologic examination revealed atypical glandular structures and interstitial inflammation.

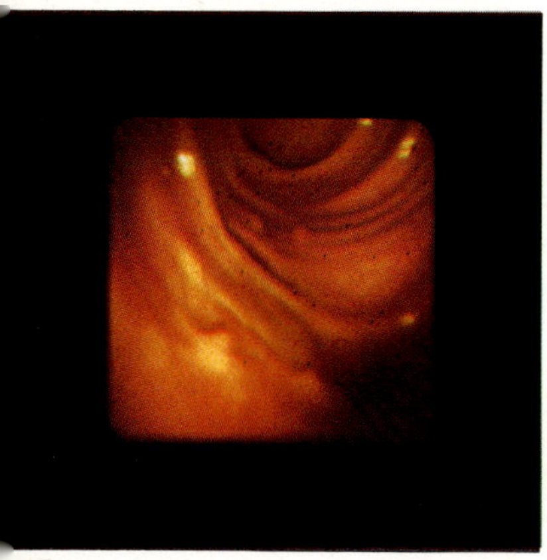

Fig. 206.
Pseudopolyps in a case of inactive idiopathic ulcerative rectocolitis. Small growths covered with smooth, shiny, pinkish mucosa stand out from the shiny, hyperaemic brittle mucosa of the descending colon.

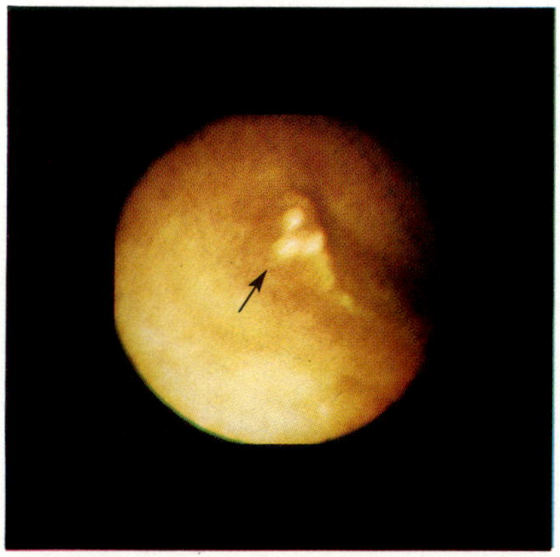

Fig. 207.
Pseudolobulated growth, the size of a grain of rice, which seems to infiltrate the surrounding mucosa in the descending colon. The mucosa is moderately hyperaemic and granular, and forms low folds.
Histologic examination revealed a nest of cancer cells. Early malignant change in a mucosa affected by long-standing idiopathic ulcerative rectocolitis was discovered accidentally in the course of endoscopy.

Crohn's disease of the colon

Fig. 208.
Barium enema. Gradual narrowing of the descending colon. The bowel wall is rigid and has an irregular contour with numerous small, round, sharply-defined filling defects. There is a localized contracture at the middle third of the transverse colon. The remaining parts of the large intestine are apparently normal.

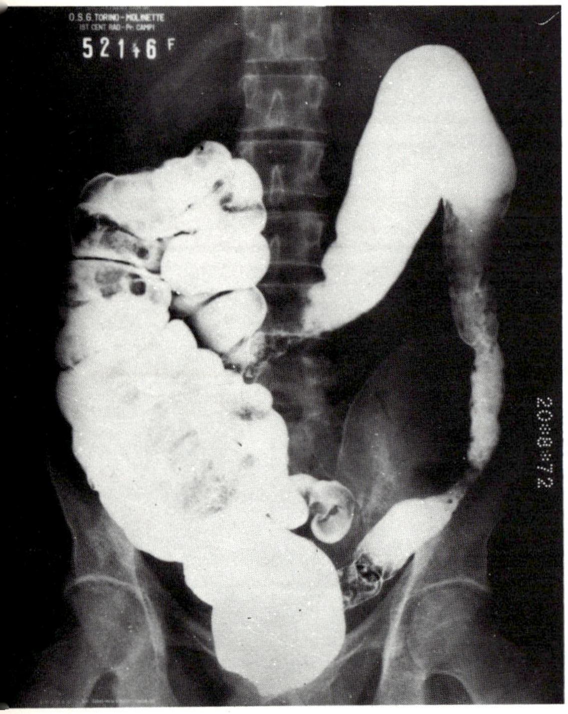

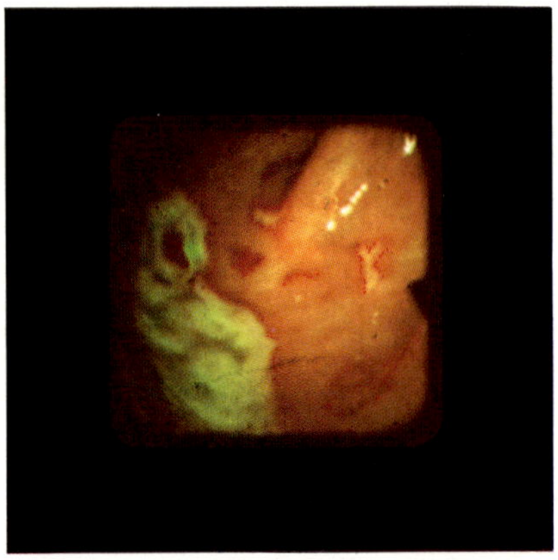

Fig. 209.
The typical early and more advanced lesions of Crohn's disease of the colon can be seen together in this figure: circumscribed hyperaemia on an apparently normal mucosa, aphtha-like ulcers, covered with fibrin, with a narrow hyperaemic border, large linear ulcer parallel with the long axis of the large bowel covered with fibrin, containing a papular lesion near its upper extremity.

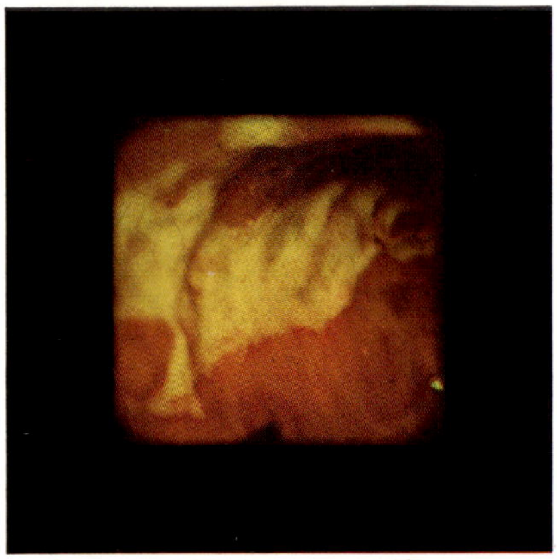

Fig. 210.
Descending colon lined with smooth pinkish mucosa with a longitudinal, furrow-like ulcer with a serpiginous course extending towards the left flexure of the colon. The borders of the ulcer are flat and bright red in colour. The bottom of the ulcer is covered with pearl-grey material (fibrinous leukocytic pseudomembranes).

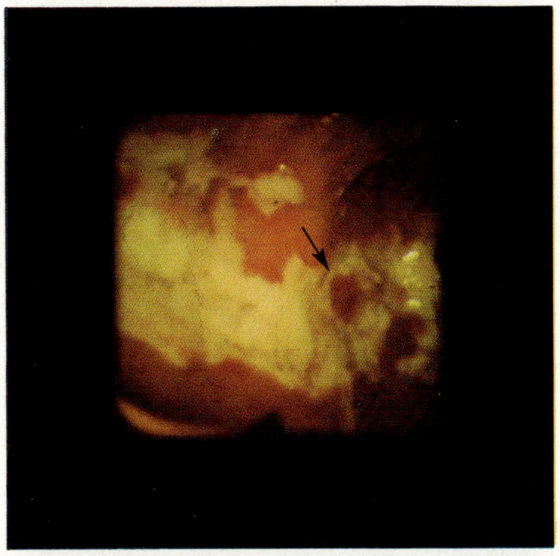

Fig. 211.
Same case. Longitudinal ulcer with a serpiginous course in which a hard, ruby-red papular growth can be seen.

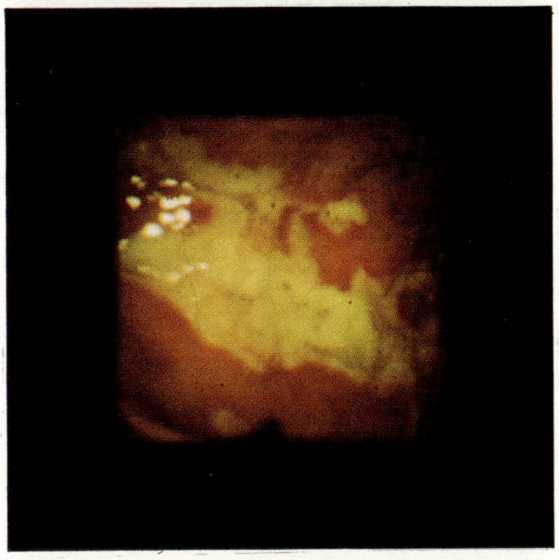

Fig. 212.
Closer view of the same case.

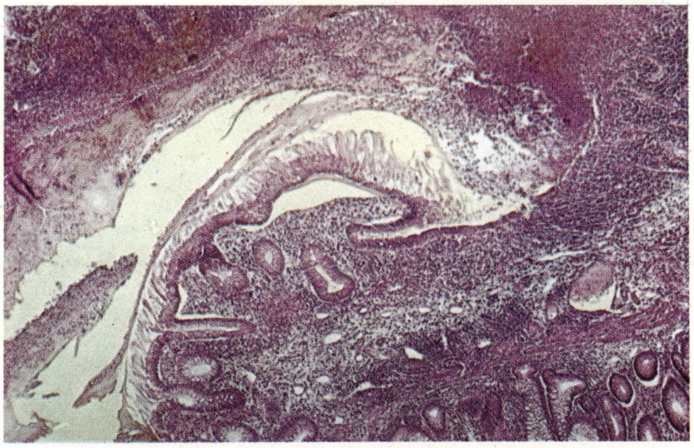

Fig. 213.
Coloscopic biopsy taken from a papular growth using the diathermy snare (big particle biopsy). Portion of ulcerated mucosa of the large intestine. Granulation tissue under a layer of exudate.

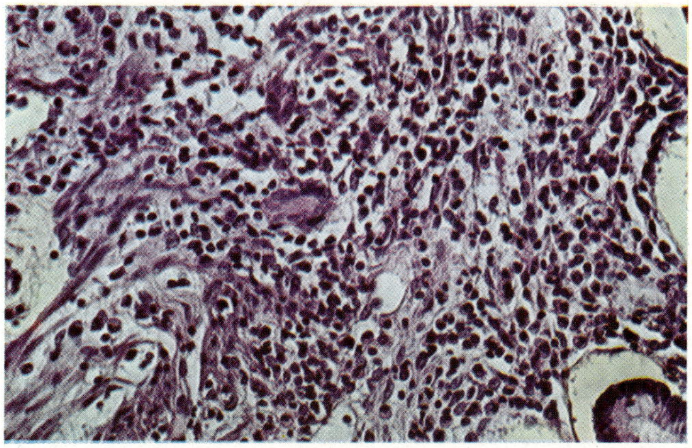

Fig. 214.
Same case. Big particle biopsy. Small granuloma (near an ulcer) characterized by a multinucleated Langhans-type giant cell, lymphocytes, histiocytes and plasma cells.

Fig. 215.
Serpiginous ulcer with hyperaemic borders running parallel to the long axis of the colon, covered with pearl-grey material from which a hard, bright-red papular lesion stands out clearly.

Fig. 216.
Same case. The mucosa is not very distensible and is apparently infiltrated on palpation with the biopsy forceps.

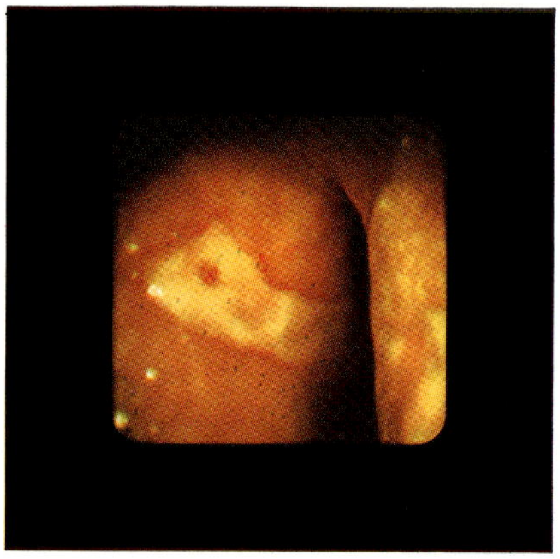

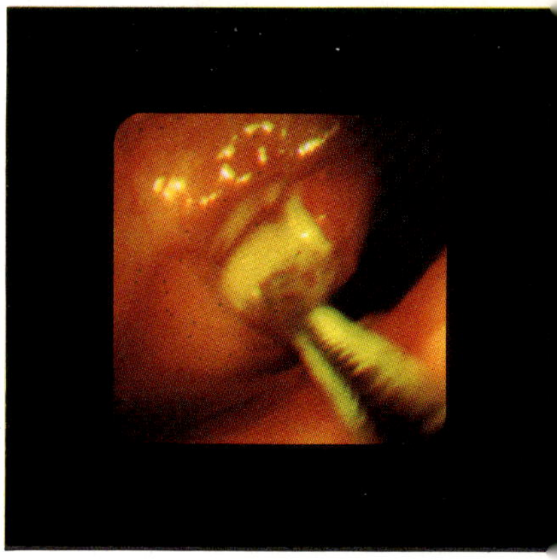

Numerous coloscopic biopsies taken from the borders of the ulcer, from the papular pseudopolypoid growth and from the adjacent mucosa revealed evidence of chronic inflammation. Histologic examination of the surgical specimen showed granulomatous colitis or Crohn's disease of the large intestine.

Fig. 217.
Barium enema. Irregular narrowing of the left colon with stiffness of the bowel wall and a multitude of small, round filling defects and scattered filling defects of larger size.
The contour of the transverse colon close to the splenic flexure is also irregular, with small ulcers on its lower aspect.

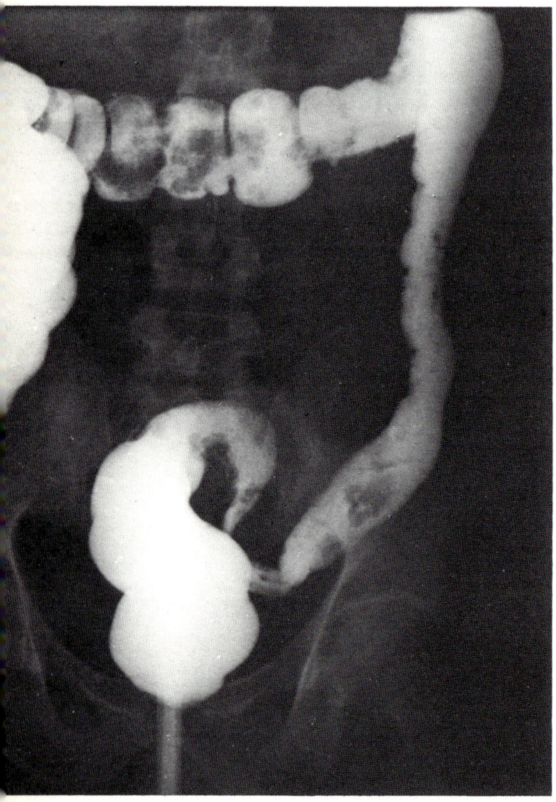

Fig. 218.
The first portion of the sigmoid colon is lined with pale, smooth, thin mucosa through which the submucosal vascular network can be seen. A sharply defined hyperaemic area is also present.

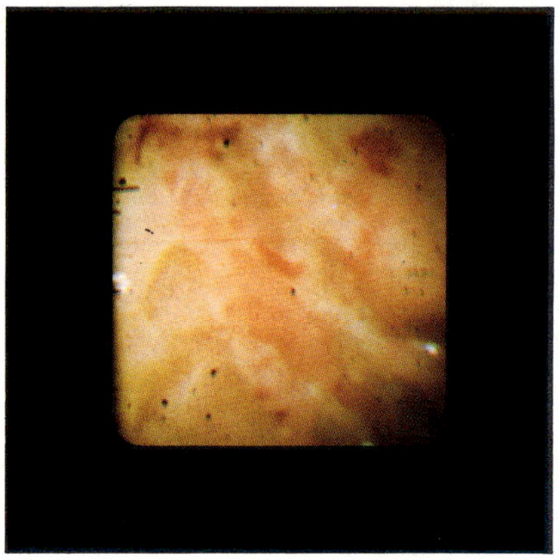

Fig. 219.
The mucosa of the second portion of the sigmoid colon shows numerous hyperaemic areas and linear ulcers with flat borders, covered with greyish material.

Fig. 220.
Aphtha-like ulcers with bright-red borders surrounded by moderately hyperaemic mucosa.

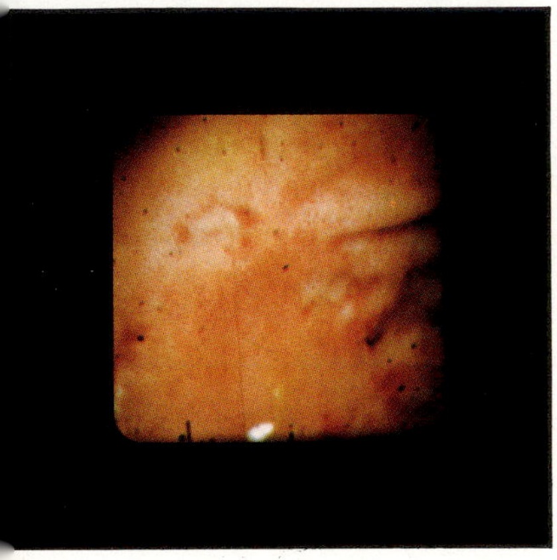

Fig. 221.
The haustrations of the descending portion of the sigmoid colon are rigid, infiltrated and considerably thickened. In the foreground there is a pseudopolypoid lesion.

Fig. 222.
Papular growths of ruby-red colour are visible at the sigmoid-descending colon junction. The mucosa forms stiff, infiltrated folds which converge towards the bowel lumen.

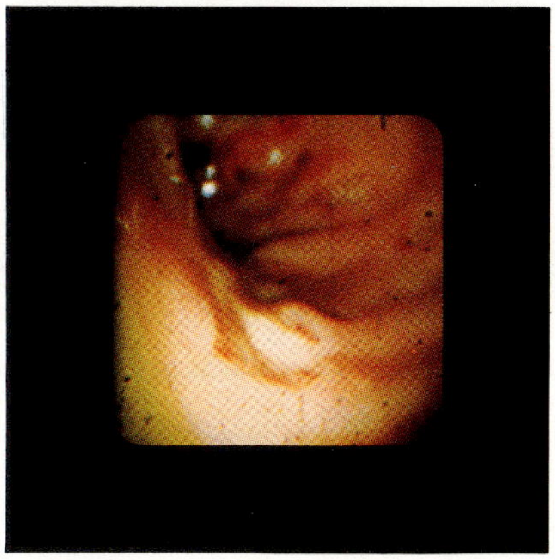

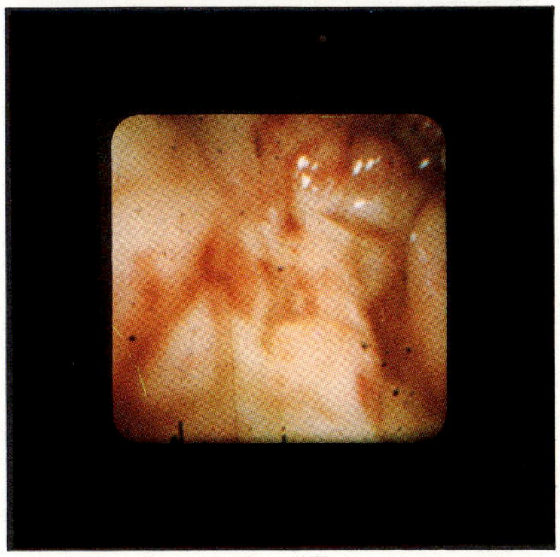

Fig. 223.
Substenosis of the distal portion of the descending colon. The bowel lumen is considerably narrowed. The adjacent mucosa is infiltrated, ulcerated, with hard, ruby-red-coloured excrescences.

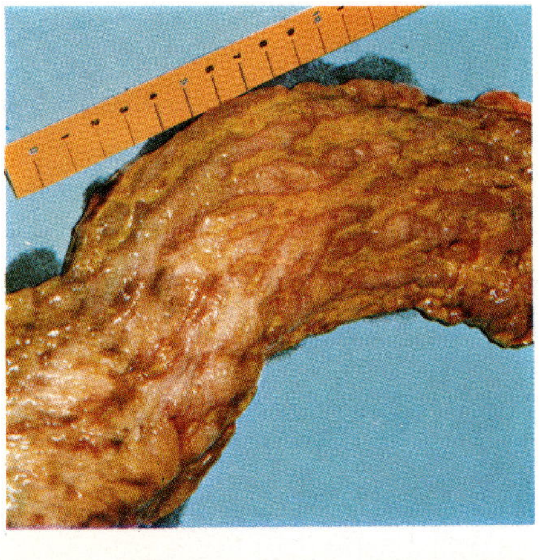

Fig. 224.
Surgical specimen. Portion of colon. The mucosa is greyish-pink in colour, with pinkish-red elevations separated by rough longitudinal furrows. The wall of the colon is more than 2 cm thick. The longitudinal furrows correspond to areas in which the mucosa has fallen and has been replaced by exudate. Histologic examination showed granulomatous colitis or Crohn's disease of the large bowel.

Fig. 225.
An aphtha-like ulcer with bright red borders, 1 mm across, stands out from the apparently normal mucosa of the left flexure.

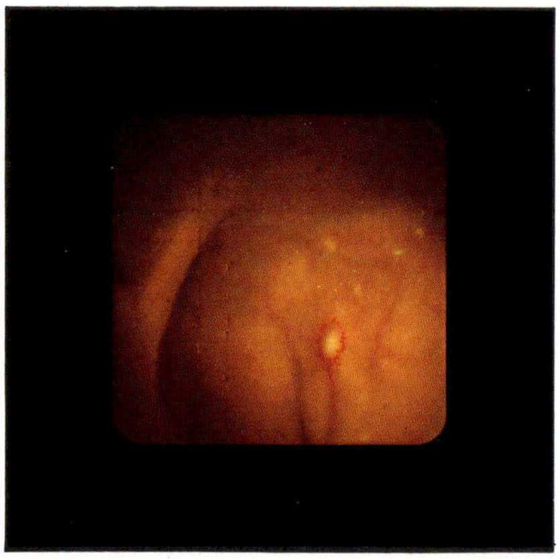

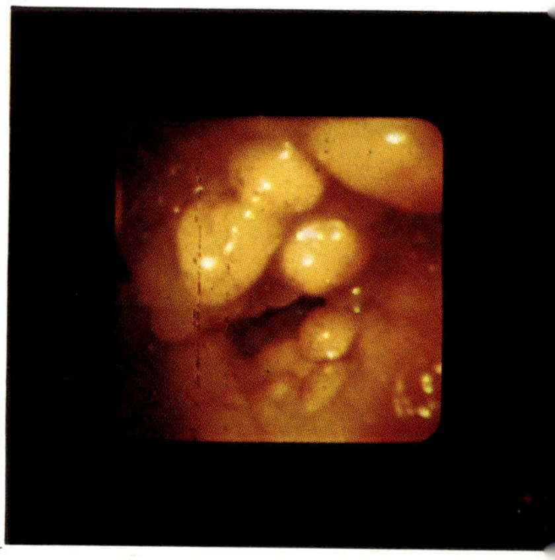

Fig. 226.
Same case. A group of hard, ruby-red papula-like polypoid lesions narrows the lumen of the transverse colon.

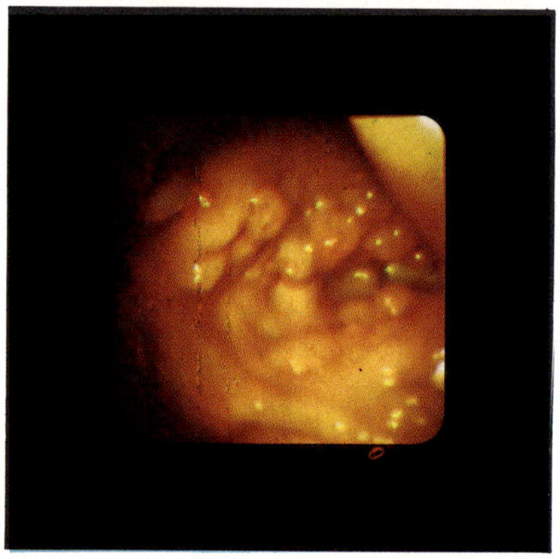

Fig. 227.
Same case. Substenosis of the middle third of the transverse colon. The bowel lumen is markedly narrowed, fixed, patent and surrounded by pseudopolypoid growths which infiltrate the bowel wall.

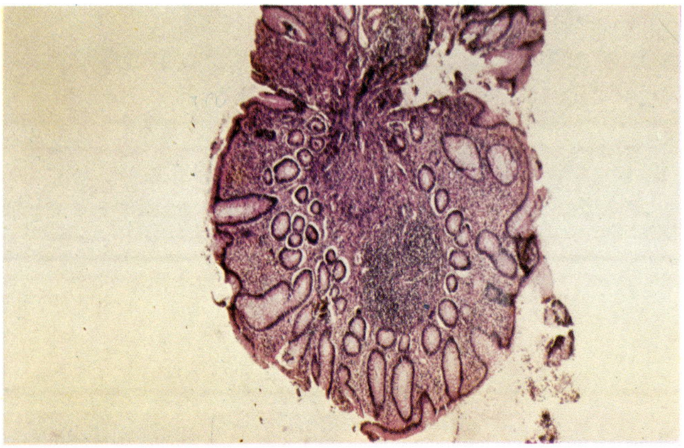

Fig. 228.
Coloscopic biopsy. Chronic aspecific inflammation.

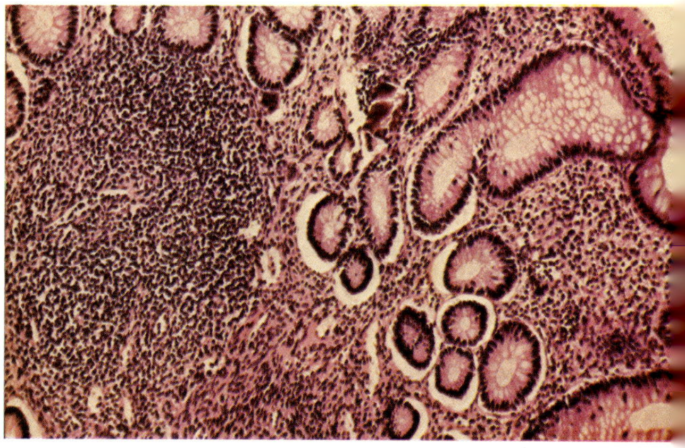

Fig. 229.
Detail of the same photomicrograph with higher magnification. Interstitial infiltration and a hyperplastic follicle.

Fig. 230.
Surgical specimen. Portion of colon. The anatomical structure of the transverse colon is upset at the site of the stenosis observed at endoscopy.

Fig. 231.
Detail of the same photograph. The wall of the transverse colon is entirely devoid of mucosa. Ulcers which infiltrate the bowel wall with intraparietal fistulization are apparent.

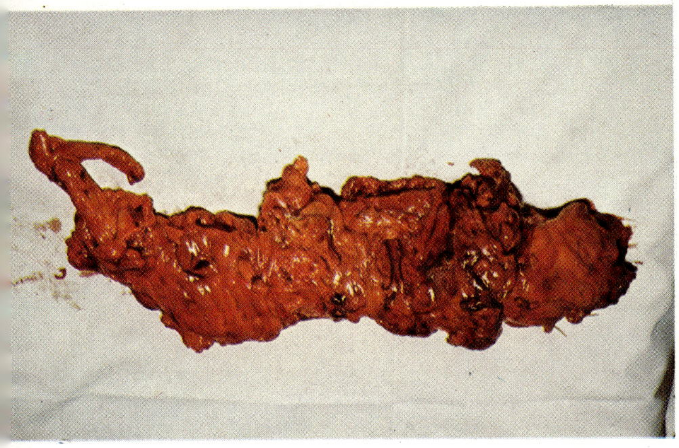

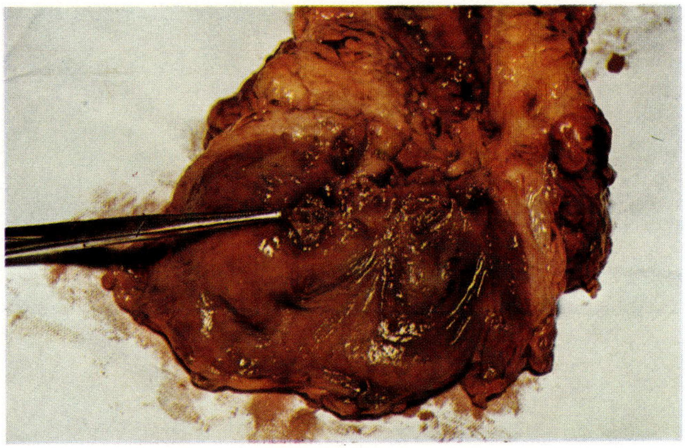

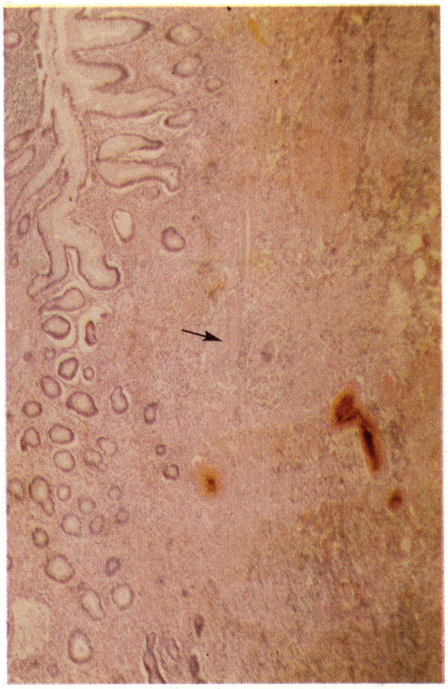

Fig. 232.
Histologic preparation of the surgical specimen. Tuberculoid granulomas.

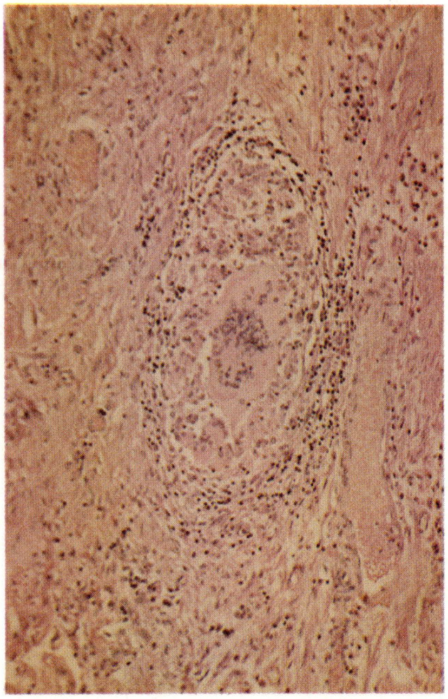

Fig. 233.
Detail of the same photomicrograph. Granuloma formed by a central multinucleated giant cell surrounded by a coat of epithelioid cells and a peripheral layer of lymphocytes.

Fig. 234.
Barium enema. The colon is tubular and markedly reduced in calibre. The bowel wall is stiff, with finely irregular contours. The descending colon shows a cobblestone pattern. The terminal ileum is also narrowed, stiff, with annular strictures and ulcer-like lesions.

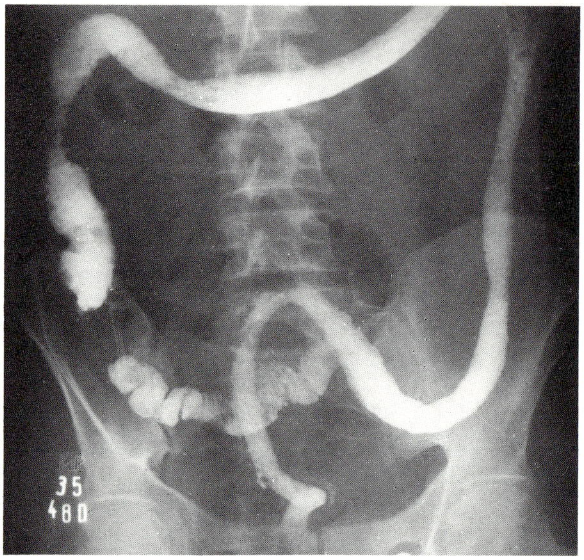

Fig. 235.
Descending portion of the sigmoid colon lined with greyish-pink mucosa raised in high, elastic folds. Specks of blood-free mucus.

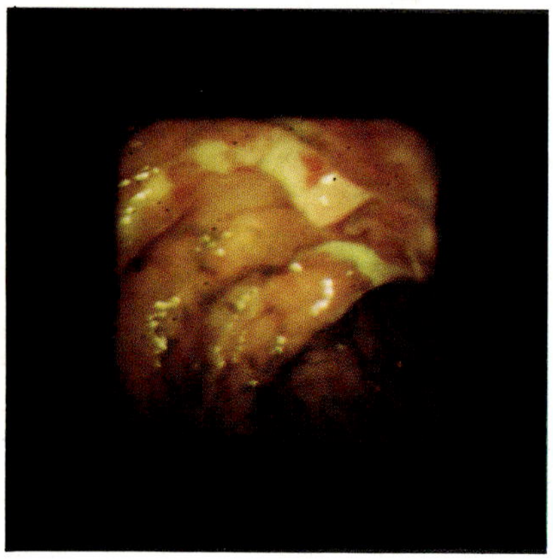

Fig. 236.
Distal portion of the descending colon lined with pinkish-red mucosa rising to form thick, apparently infiltrated folds of increased consistency, with shallow ulcers.

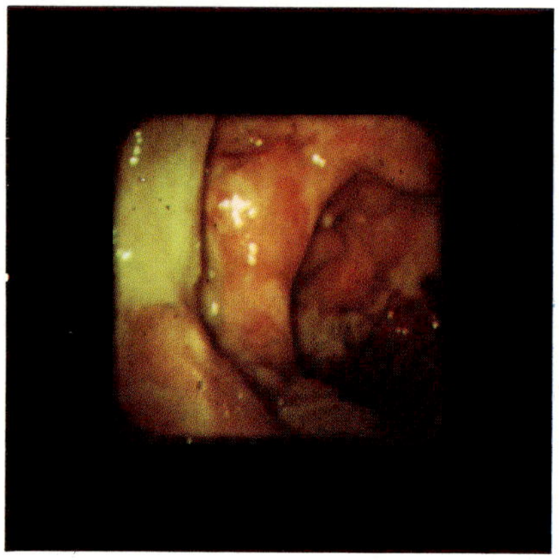

Fig. 237.
The descending colon has a characteristic cobblestone-like appearance due to numerous papular lesions. The bowel wall is stiff and infiltrated. Peristalsis is absent. The haustral folds are no longer recognizable.

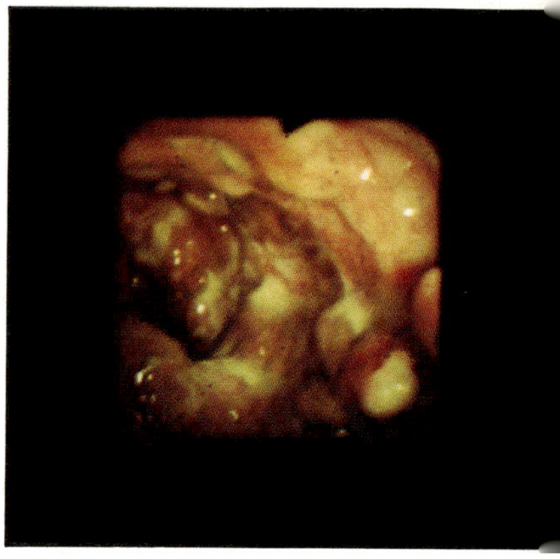

ig. 238.
ard, ruby-red-coloured
seudopoylps (papulae)
 the foreground.

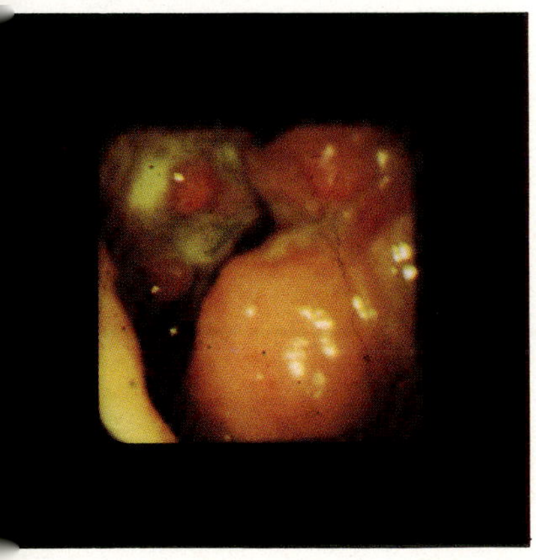

Fig. 239.
Extensive ulceration surrounding a large pseudopolypoid growth.

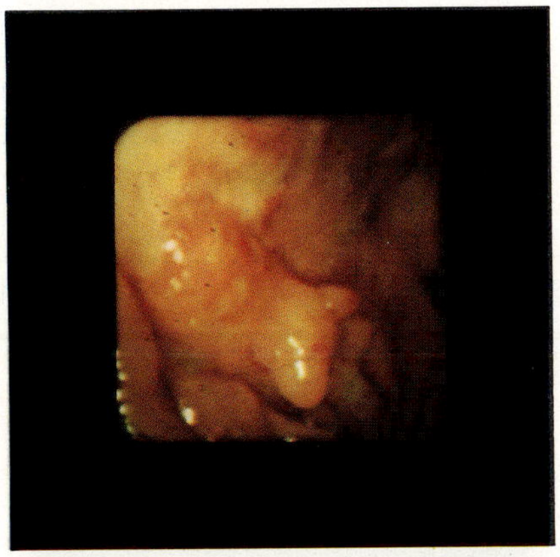

Fig. 240.
Coloscopic biopsy on the pseudopolyps (papulae) in the foreground.

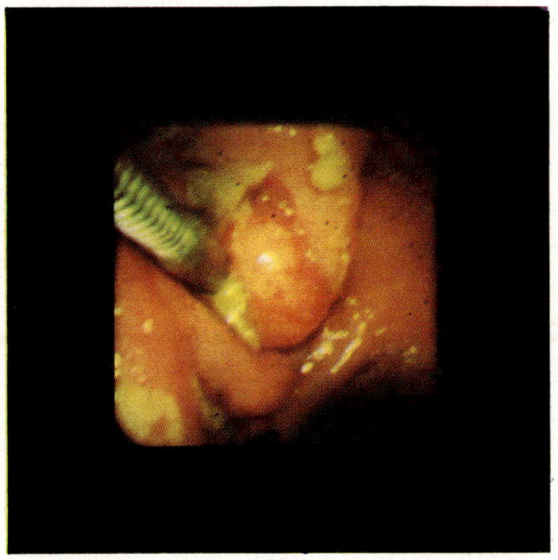

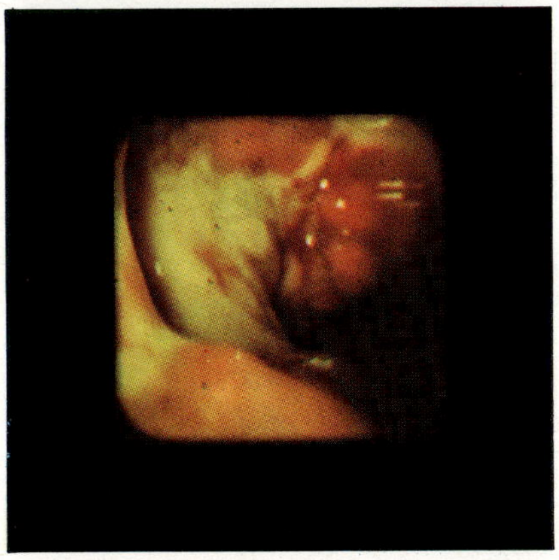

Fig. 241.
Bleeding after coloscopic biopsy is slight.

Figs. 242 e 243.
Hard papular lesions bordered by longitudinal ulcers on the mucosa of the transverse colon which thus also shows a cobblestone-like appearance.

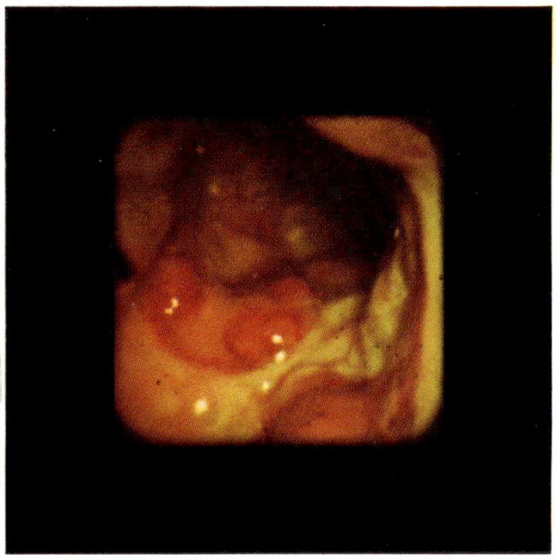

Fig. 243.

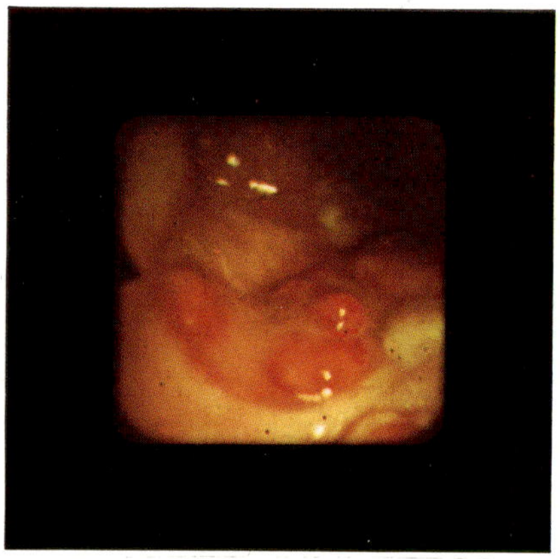

Fig. 244.
Incomplete stenosis of the first portion of the ascending colon which prevents further advancement of the fibrescope.

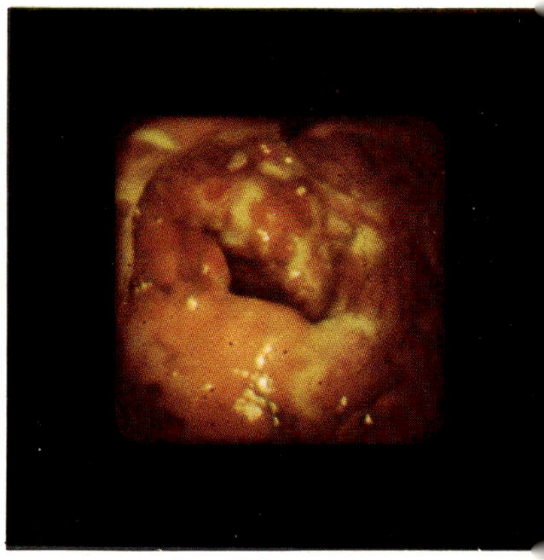

Terminal ileitis

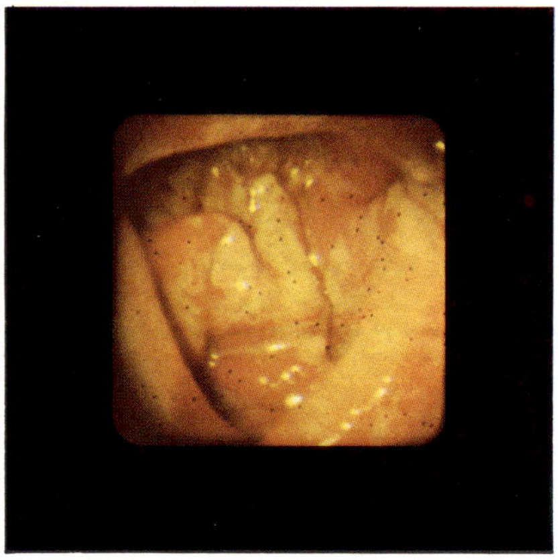

Fig. 247.
View of the last ileal loop through the ileocecal papilla. The ileum is lined with pinkish-red mucosa which is thickened by oedema and forms folds which on palpation with the biopsy forceps prove not very elastic.
Shallow ulcers with ill-defined borders are also visible.

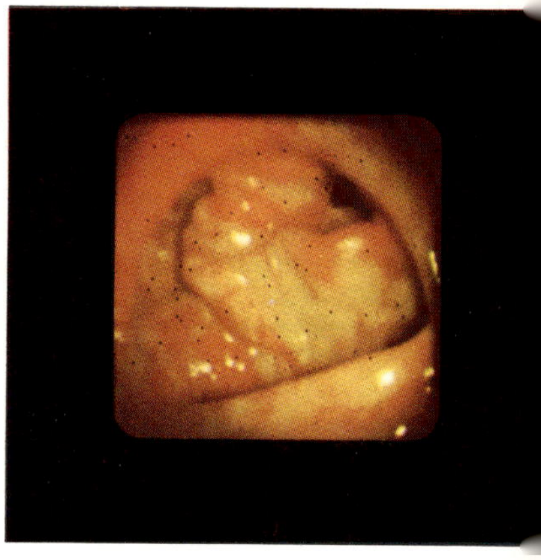

Fig. 248.
Closer view of the same case. The severe oedema and infiltration of the mucosa narrow the lumen of the last ileal loop and make advancement of the fibrescope difficult and dangerous. The bowel wall cannot be distended even with high insufflation pressures. Histologic examination of the surgical specimen revealed granulomatous terminal ileitis.

Fig. 245.
Crohn's disease of the colon. Coloscopic biopsy: epithelioid granuloma with central giant cell.

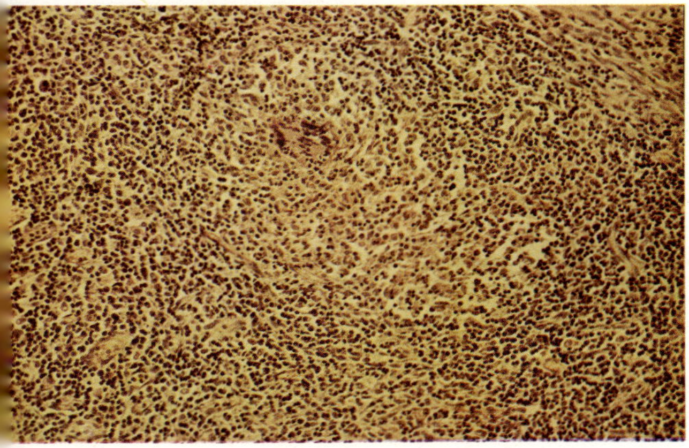

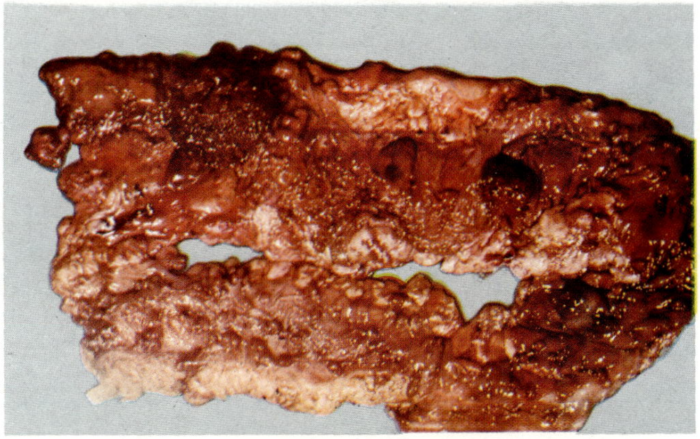

Fig. 246.
Surgical specimen. Portion of large intestine about 110 cm long. The bowel wall is thickened (about 2 cm) by connective tissue. The outer surface of the bowel has a granular appearance. The mucosa is greyish-pink in colour and uneven on account of the presence of pinkish-red excrescences of different sizes, bordered by longitudinal furrows covered with mucus and fibrin. Fissures on the bowel wall reach the subserous tissues.

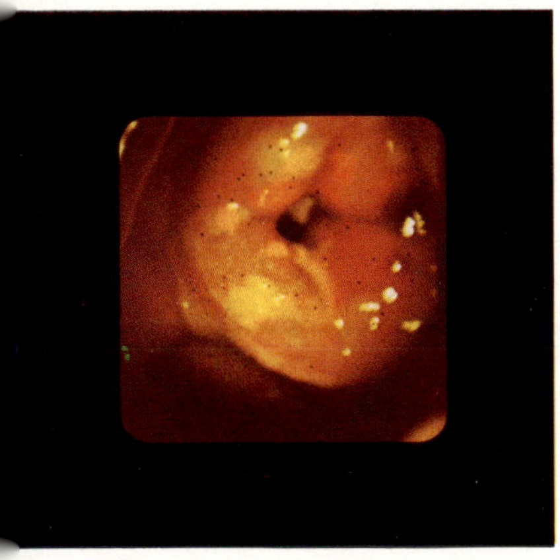

Fig. 249.
The lumen of the last ileal loop is clearly narrowed, with stiff, infiltrated walls. The mucosa is intensely congested, ulcerated, and infiltrated. Histologic examination of the surgical specimen revealed granulomatous terminal ileitis.

distinctive endoscopic features of Idiopathi

IDIOPATHIC
ULCERATIVE
RECTOCOLITIS

- Diffuse oedema and congestion—shiny mucosa
- Bleeding on contact—abundant after biopsy
- Widespread ulcerations:
  - superficial erosions
  - microulcers
  - linear or stellate ulcers
  - punched-out ulcers
- Oozing hemorrhages
- Friable inflammatory polyps
- Granular mucosa
- Bowel wall not infiltrated
- Haustra few or absent
- Little or no motility of bowel wall
- Abundant secretion (serous, mucofibrinous, purulent)

ROHN'S DISEASE	— Sharply defined areas of hyperemia
	— No bleeding on contact—little after biopsy
	— Aphtha-like ulcers, scattered or grouped (diameter 2-5 mm)
	— Normal mucosa surrounding lesions
	— Flat longitudinal ulcers with red borders—bottom covered with pearl-grey material
	— No oozing hemorrhages
	— Hard papular growths
	— "Cobblestone" mucosa
	— Rigid bowel wall
	— Little or no secretion
	— Fistulization
	— Stenosis

Pseudomembranous colitis

Fig. 250.
Barium enema. X-ray taken after partial spontaneous evacuation of the contrast. The mucosal folds are increased in height and thickness and, especially in the proximal section of the colon, form a wide-meshed network outlining roundish, pseudolacunar areas. The descending colon shows evidence of oedema of the mucosa with uneven adhesion of the barium to the bowel wall.

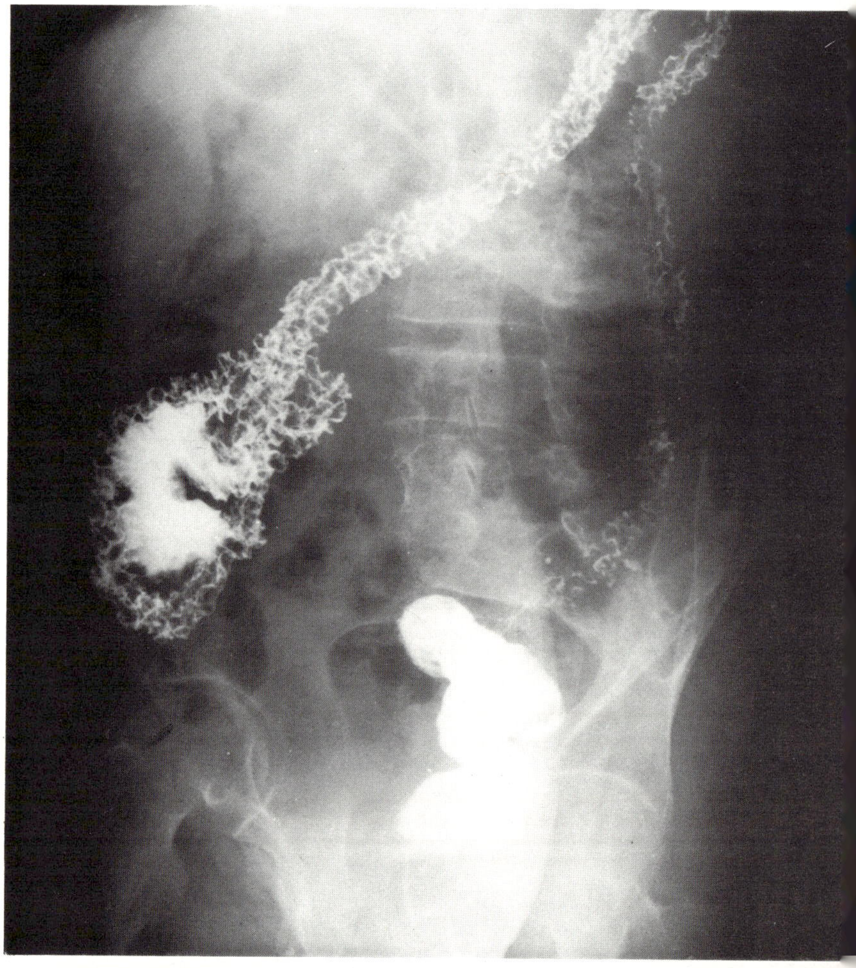

Figs. 251, 252 & 253.
Pseudomembranous colitis. Endoscopic view of the sigmoid colon (Fig. 251), descending colon (Fig. 252) and transverse colon (Fig. 252). The distensibility of the colon by insufflation is reduced. The bowel wall is lined with intensely congested and oedematous mucosa. When the patches are removed with the aid of the biopsy forceps the underlying mucosa appears to be ulcerated and oozes blood.
(Courtesy of prof. G. Verme, Torino)

Fig. 251.

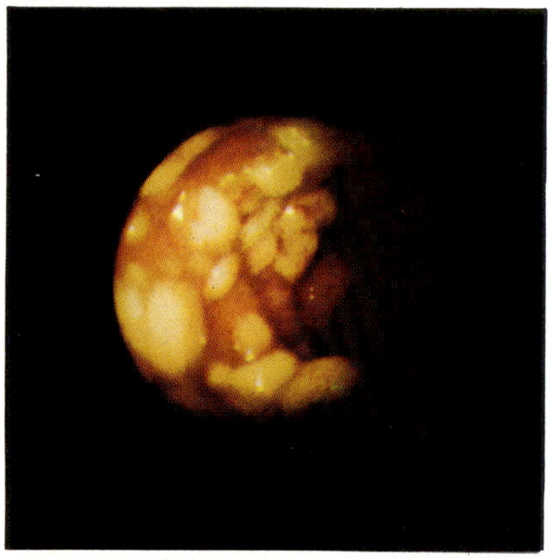

Fig. 252.

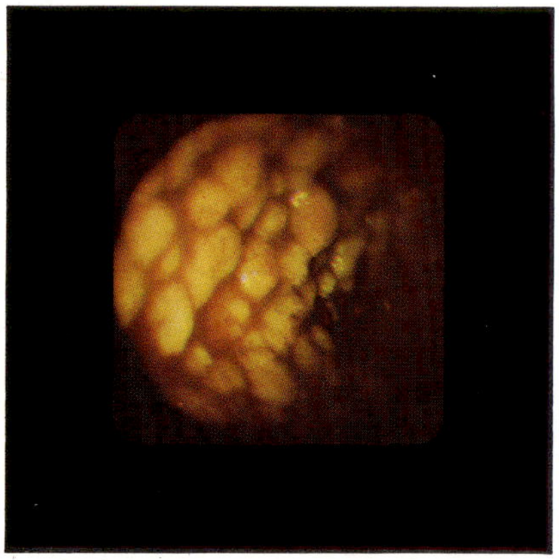

Fig. 253.

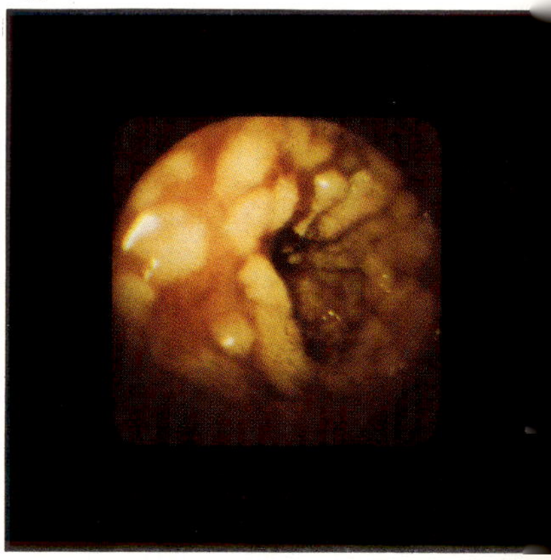

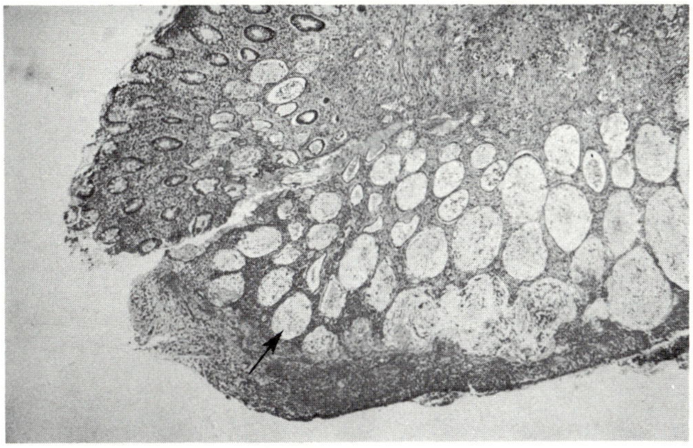

Fig. 254.
Coloscopic biopsy. Histologic preparation. Haematoxylin-eosin stain. Ulcerated mucosa of the large intestine. The glands are swollen by a necrotic fibrinoid material containing many neutrophilic leukocytes. Note the sharply-defined boundary between the apparently normal mucosa (top left) and the dilated glands (bottom).

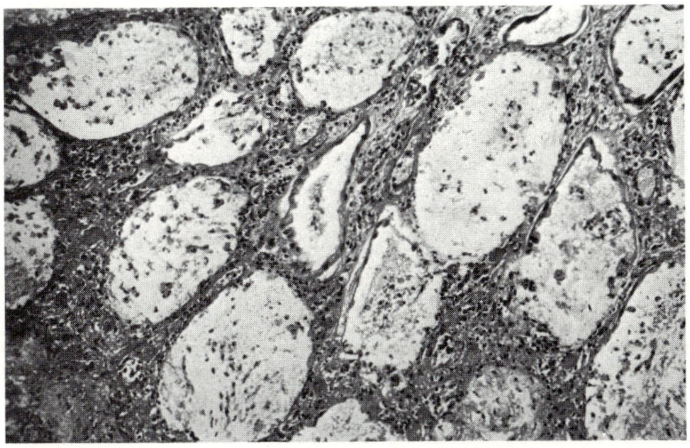

Fig. 255.
Detail of the same preparation at higher magnification. The epithelial layer of the glands is flattened or destroyed. An aspecific inflammatory process can be seen in the interstitial tissue with a predominance of polymorphonuclear leukocytes. The histologic picture is typical of pseudomembranous colitis.
(Courtesy of prof. G. Verme, Torino)

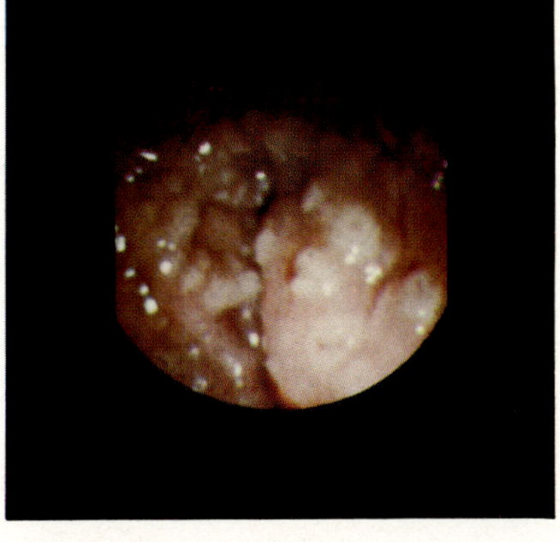

Fig. 256.
Pseudomembranous colitis. The walls of the sigmoid colon are lined with oedematous and hyperaemic mucosa with numerous lardaceous patches showing the same endoscopic features as the previous case.

Sigmoiditis aspecific

Fig. 257.
Barium enema. The iliac-pelvic portion of the colon is apparently shortened on account of spastic contractures of the sigmoid colon. The haustrations of the sigmoid colon are irregular, thin, close together, with a saw-like appearance and small marginal projections.

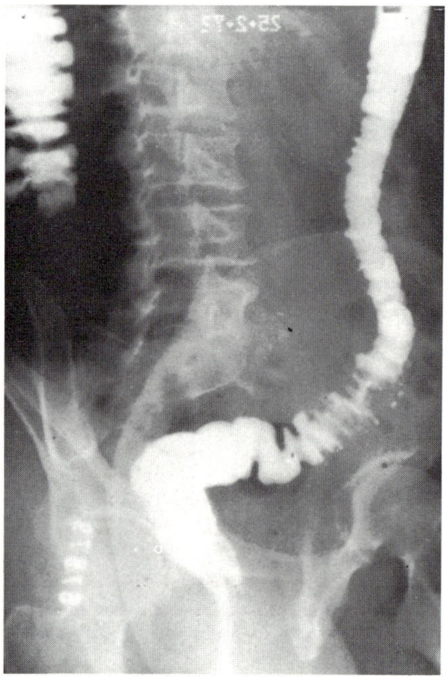

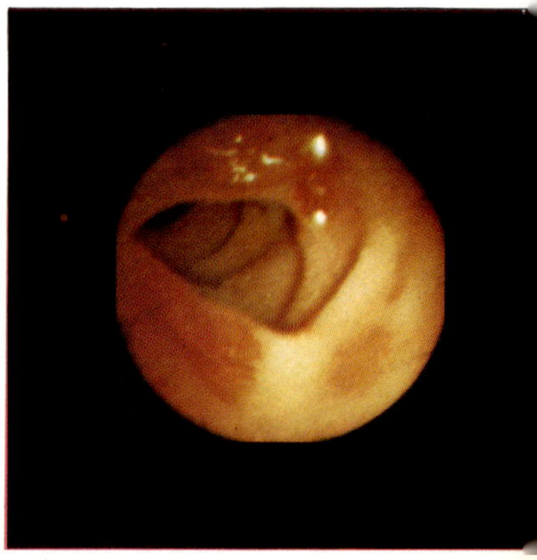

Fig. 258.
Aspecific sigmoiditis. The descending portion of the sigmoid colon is lined with greyish-pink mucosa. There are confluent patches of oedema and congestion. The folds are well-developed, oedematous and elastic. All the mucosa is brittle and bleeds easily.

Figs. 259, 260 & 261.
Aspecific sigmoiditis. The bowel wall of the second portion of the sigmoid colon is normally distensible. The mucosa is everywhere pinkish-red in colour, smooth, shiny and thin so that the underlying vascular networks show up through it.

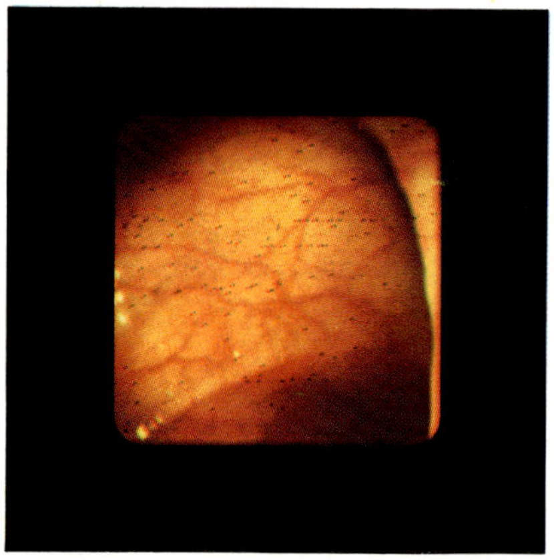

Fig. 259.

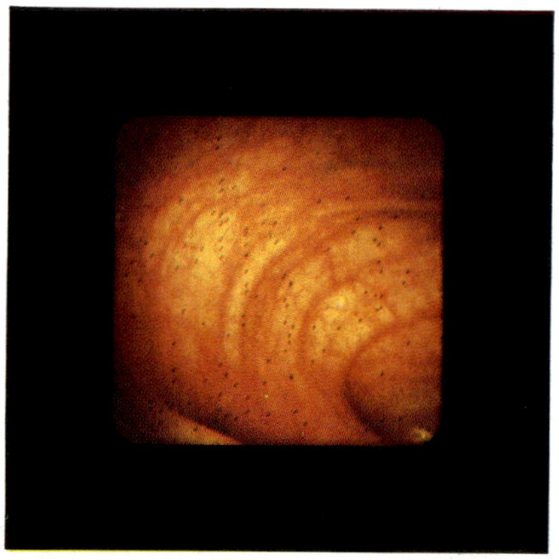

Fig. 260.

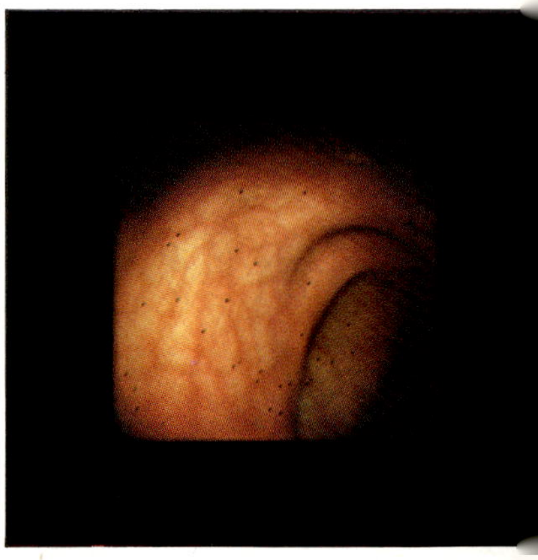

Fig. 261.

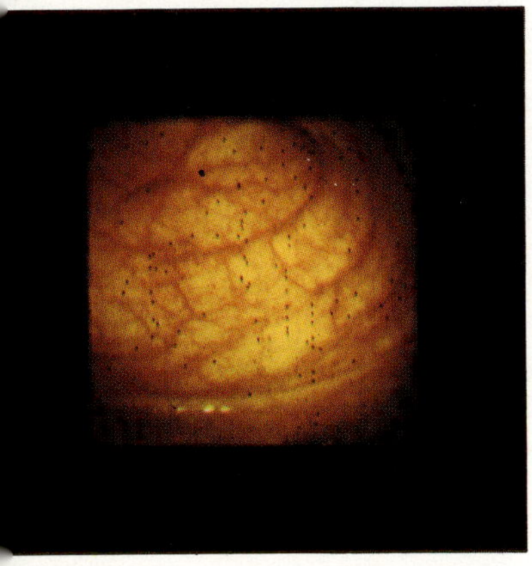

ɡ. 262.

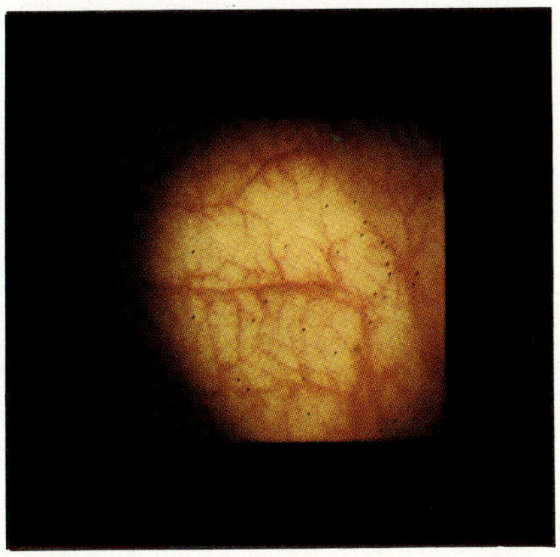

Fig. 263.

Figs. 262 & 263.
Radiation sigmoiditis. The folds of the second portion of the sigmoid colon are flattened. The mucosa is thin. The underlying vascular networks are congested.

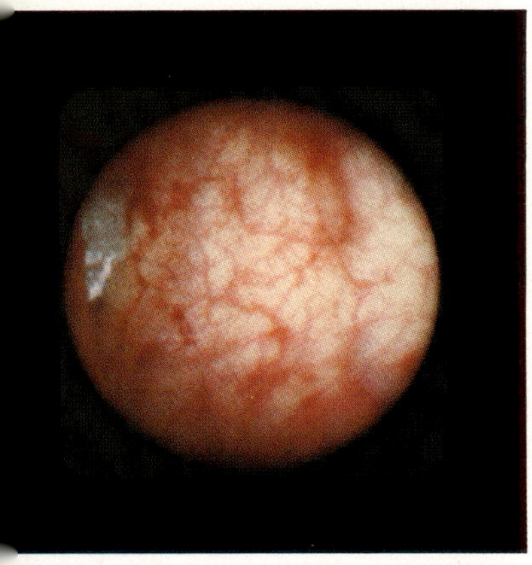

ɡ. 264.

Fig. 264.
Radiation sigmoiditis. The mucosa of the sigmoid colon is greyish-pink, smooth, shiny and thin with a very visible vascular network. A bloody serous transudate oozes from the mucosa.

Figs. 265, 266 & 267.
Incomplete stenosis of the sigmoid colon following radiation treatment. Marked narrowing of the lumen of the middle and distal portion of the sigmoid colon due to stricturing by perivisceritis which makes further progression of the fibrescope impossible. The bowel wall is indistensible. The mucosa is hyperaemic, thin, smooth and shiny. The folds are distorted and converge towards the substenosis.

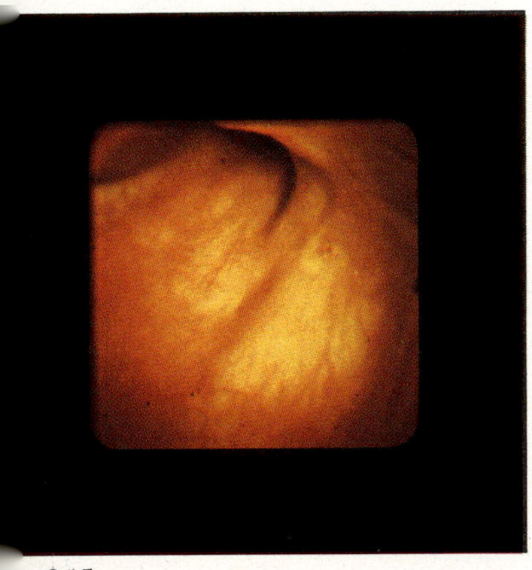

:. 265.

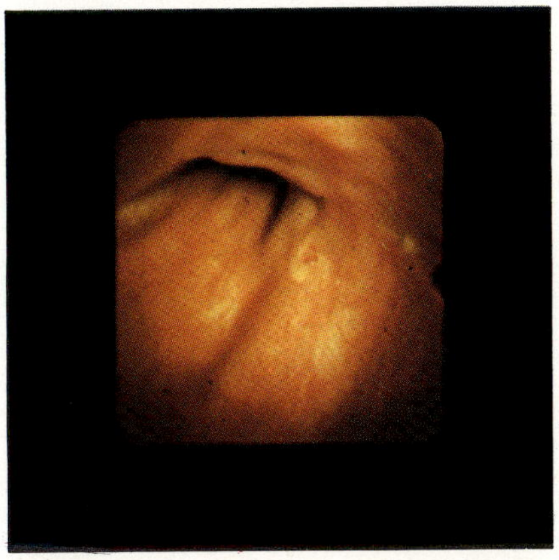

Fig. 266.

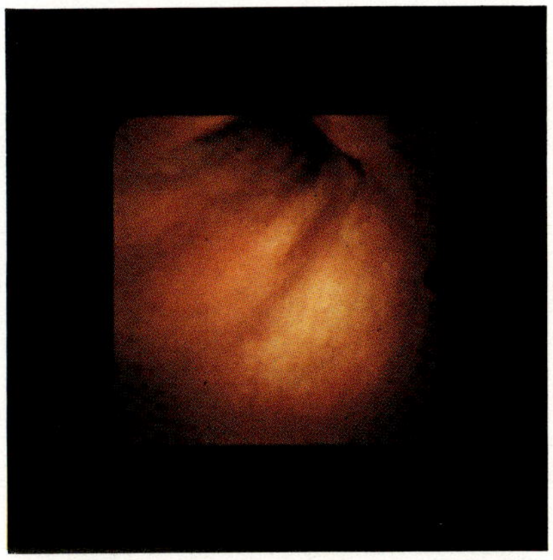

Fig. 267.

Polyps

yperplastic polyps

ʏperplastic polyps are frequently encountered during loscopy, especially in old persons. They are usually small, ssile, smooth-surfaced, and covered with greyish-pink mucosa ıich is often indistinguishable from the surrounding mucosa. ıey are very common in the sigmoid colon, not rare in other rts of the large bowel, and are almost always multiple. On lpation with the biopsy forceps they appear soft and easy to netrate.

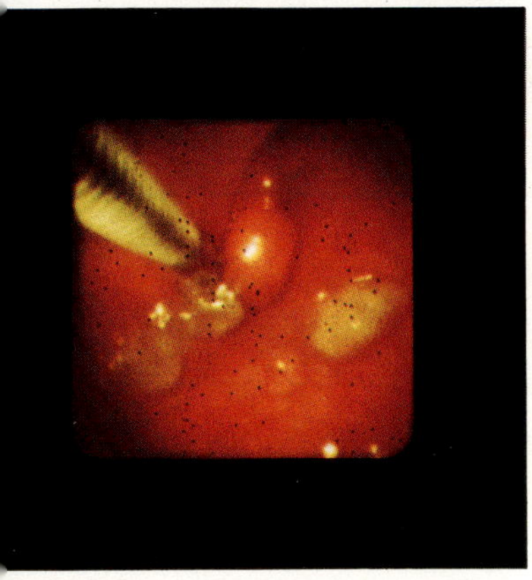

;. 268.
ıooth-surfaced, sessile crescence, 3-4 mm in size, ʋered with pink, smooth, shiny ıcosa on the wall of the sigmoid on on a background of eper-coloured mucosa. Specks of ıcus on the bowel wall. A ıpsy was taken at the base of e growth. Histologic diagnosis: hyperplastic polyp.

ʒ. 269.
me case. Another sessile polyp th the same aspect in the ınsverse colon.

Fig. 270.
Yet another sessile polyp, of identical aspect, in the ascending colon.

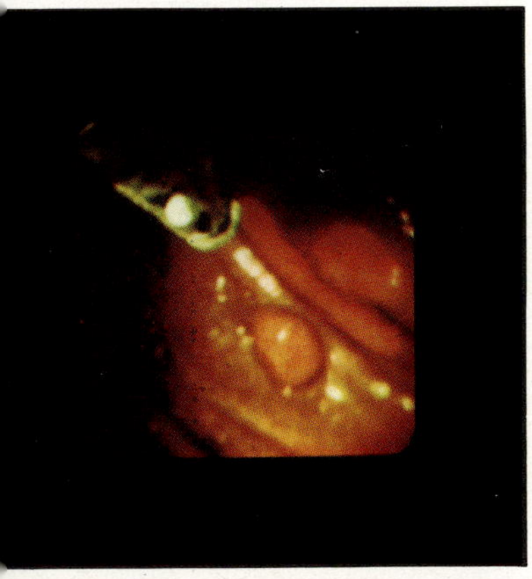

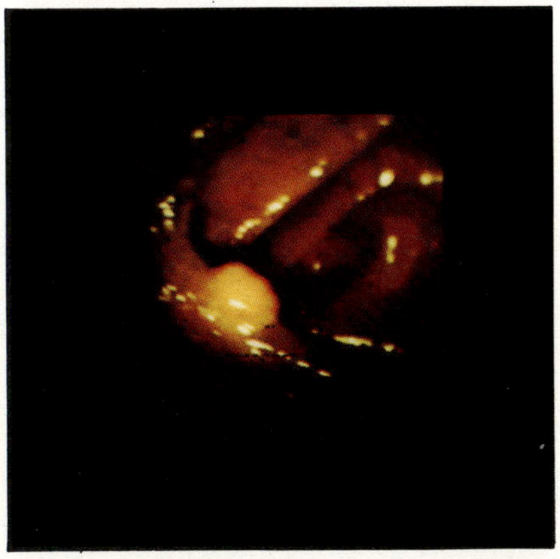

adenomatous polyps

Fig. 271.
Spherical, sessile polyp about 2 mm in size on the mucosa of the sigmoid colon. Its surface is smooth and covered with pinkish mucosa. Histologic diagnosis: adenomatous polyp.

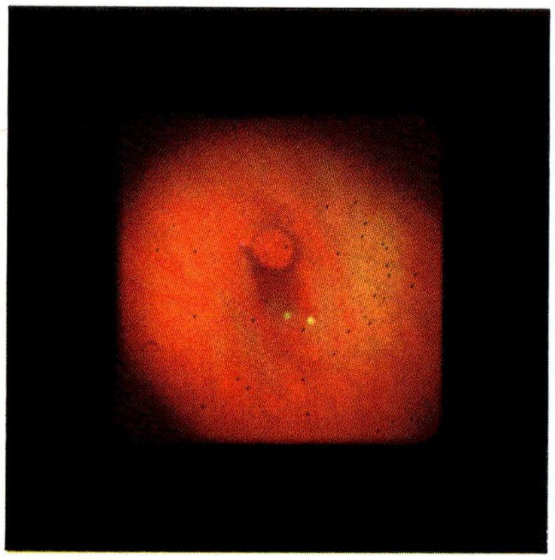

Fig. 272.
Smooth-surfaced, sessile polypoid growth about 3 mm in size, covered with pinkish mucosa, in the distal portion of the descending colon. On palpation with the biopsy forceps the polyp is hard and elastic. Specks of mucus are visible around its base. Histologic diagnosis: adenomatous polyp.

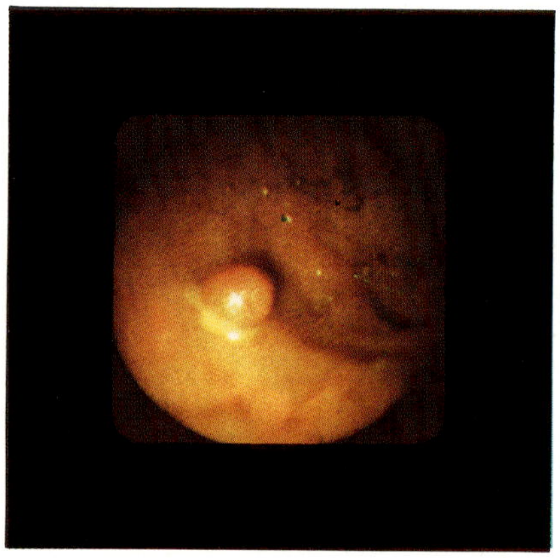

Fig. 273.
Sessile, pseudolobulated polypoid growth on the mucosa of the descending colon. Its surface is smooth and covered with pinkish-red mucosa. The polyp is hard and elastic and escapes from the clutch of the jaws of the biopsy forceps. The surrounding mucosa and semilunar fold are elastic.

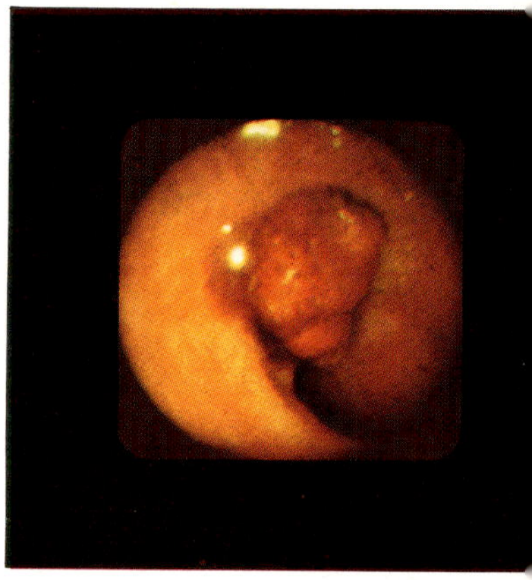

Fig. 274.
Adenomatous polyp (histologic diagnosis) of the descending colon. Its surface is smooth and covered with shiny, greyish-pink mucosa. The haustral folds are normal.

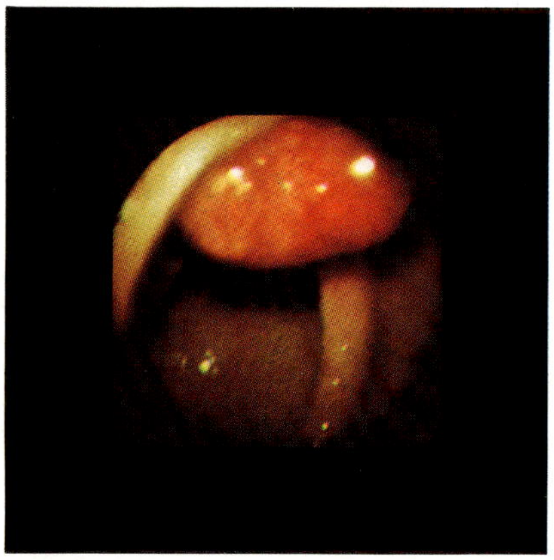

Fig. 275.
Pedunculated polyp in the distal portion of the sigmoid colon. The stalk is 1.5 cm long and is covered with smooth, greyish-pink mucosa. The base of the stalk is not infiltrated. The tip of the polyp is semispherical and is covered with bright red, shiny, smooth mucosa. Histologic diagnosis: adenomatous polyp.

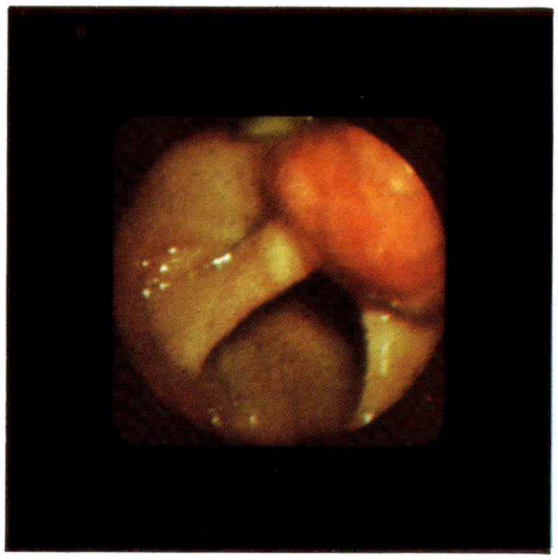

Fig. 276.
Wide-based smooth-surfaced, hard, elastic polypoid growth covered with pinkish, shiny mucosa on a fold of the descending colon. Histologic diagnosis: adenomatous polyp.

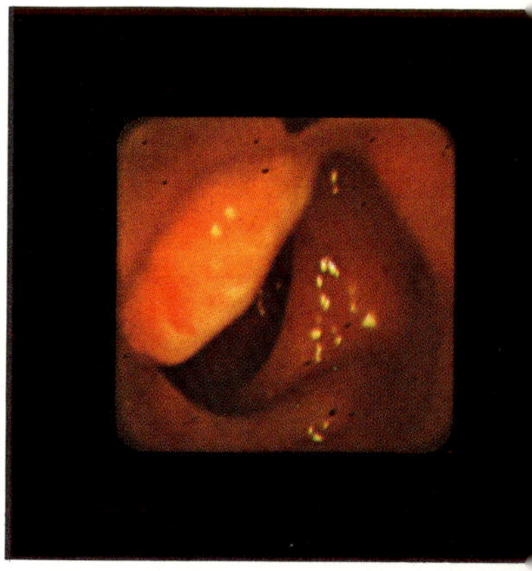

denomatous polyps are a common finding at coloscopy. They ay be single, multiple or diffuse, sessile or pedunculated (long · short stalk), of various aspect, with or without seudolobulations or true lobulations, and of various size, om 0.5 to 8 cm in diameter. Their surface is smooth and overed with greyish-pink, smooth, shiny mucosa. hey appear to be tough and elastic in consistency when uched with the biopsy forceps and easily escape from the utch of its jaws.

g. 277.
lenomatous polyp. Spherical, ssile, hard, elastic polypoid owth covered with pinkish, smooth, iny mucosa on the summit of a ld of the descending colon.

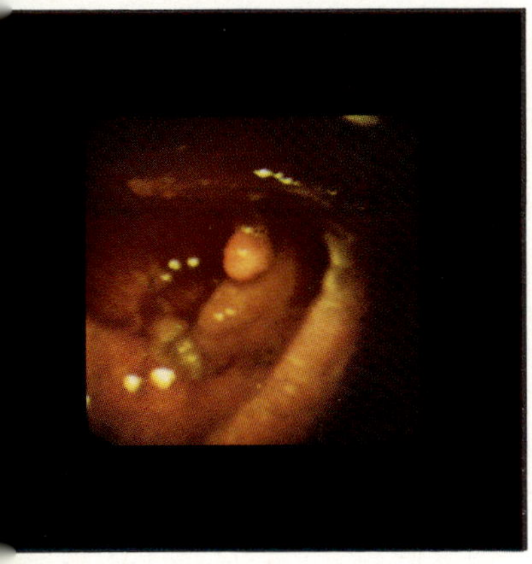

Fig. 278.
Sessile polyp the size of a hazel nut in the ascending colon. The nearby mucosa is greyish-pink, smooth and shiny, with a regular vascular network. The bowel wall is normally distensible on insufflation. Histologic diagnosis: adenomatous polyp.

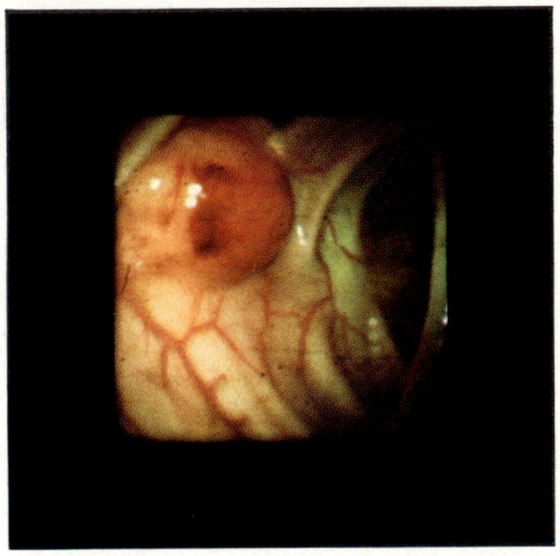

;. 279.
lenomatous polyp in a case of opathic ulcerative rectocolitis. A all, hard, elastic, brittle sessile owth the size of a grain of rice, vered with pinkish, smooth, ny mucosa stands out against intensely hyperaemic, erated, bleeding mucosa of the on involved by the active lammatory process. Histologic gnosis: Adenomatous polyp. ourtesy of prof. A. Montori, me.)

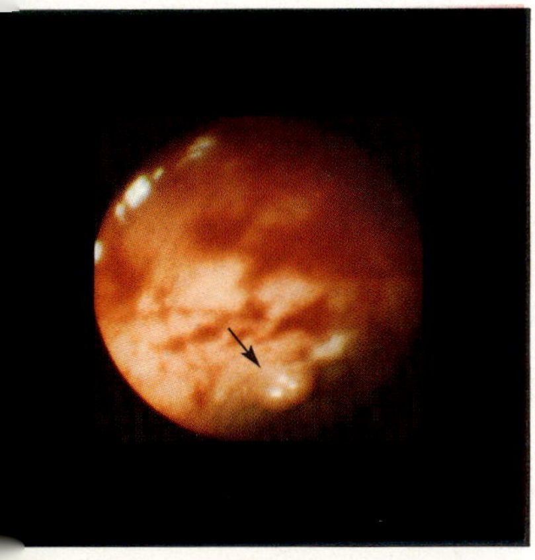

Fig. 280.
Long-stalked polyp with pseudolobulated tip covered with pinkish-red, smooth, shiny mucosa. The polyp is hard and elastic on palpation with the biopsy forceps.

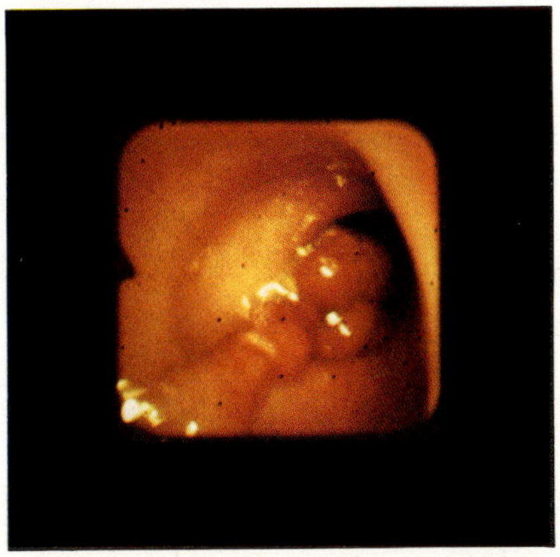

Fig. 281.
Sessile polypoid growth with a lobulated surface covered with greyish-pink, smooth, shiny mucosa. The nearby folds are not infiltrated. The polyp is hard and elastic. Histologic diagnosis: adenomatous polyp.

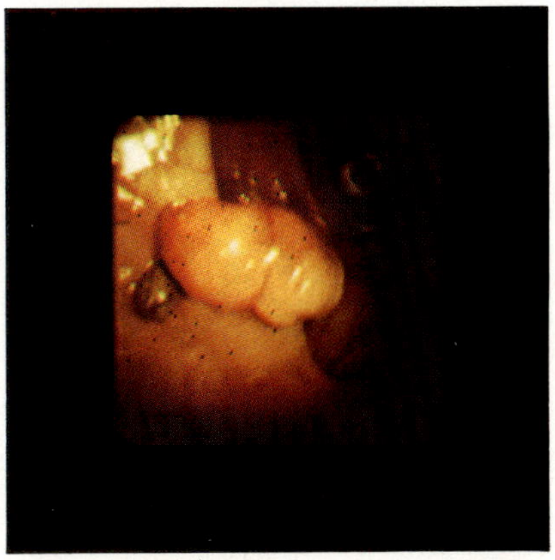

Fig. 282.
The descending colon is lined with smooth, greyish-pink mucosa which forms folds of normal thickness. A new growth, partly covered with mucus and blood-free faeces, is hidden between the haustrations.

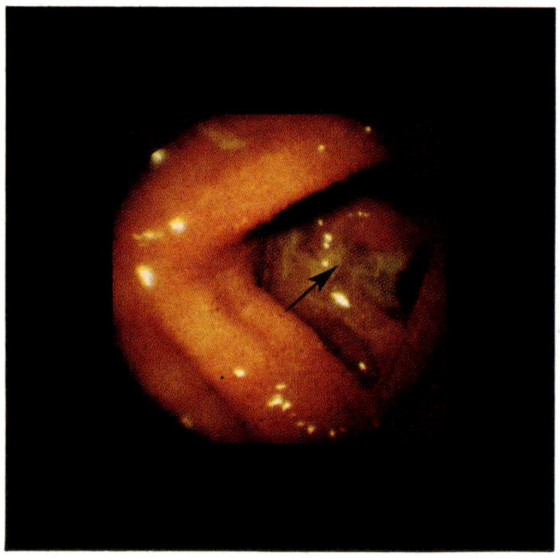

Fig. 283.
After washing out with water a short-stalked polypoid growth is visible. Histologic diagnosis: adenomatous polyp.

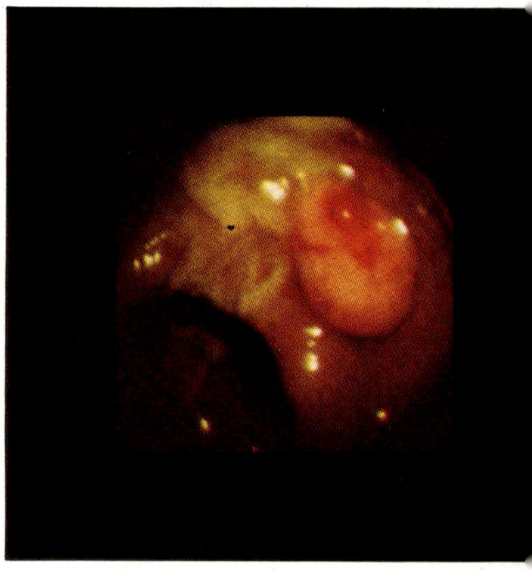

gs. 284, 285 & 286.
long-stalked polypoid growth
osses the lumen of the descending
lon. The tip of the polyp
pseudolobulated, bright red in
lour, with a finely uneven
rface (figs. 285 & 286), and is
dden by and adheres to the
wel wall opposite to the base of
stalk. Histologic diagnosis:
enomatous polyp with atypical
ndular features.

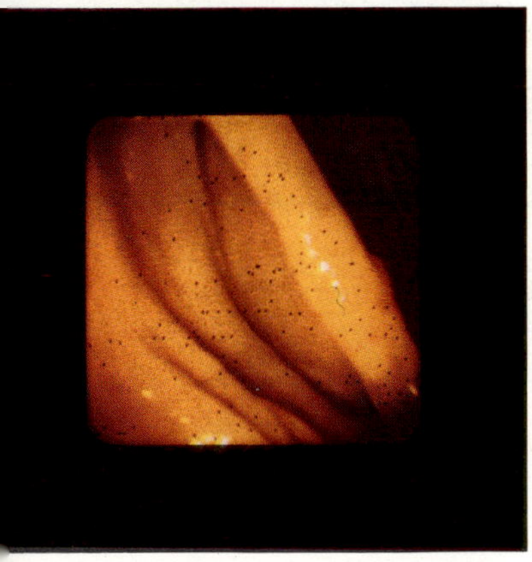

Fig. 285.

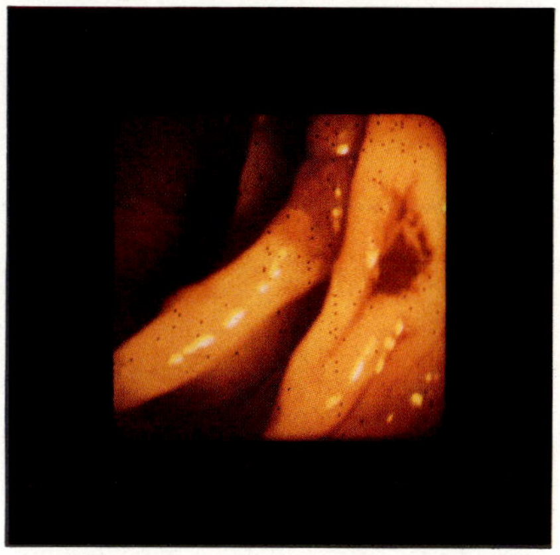

Fig. 286.

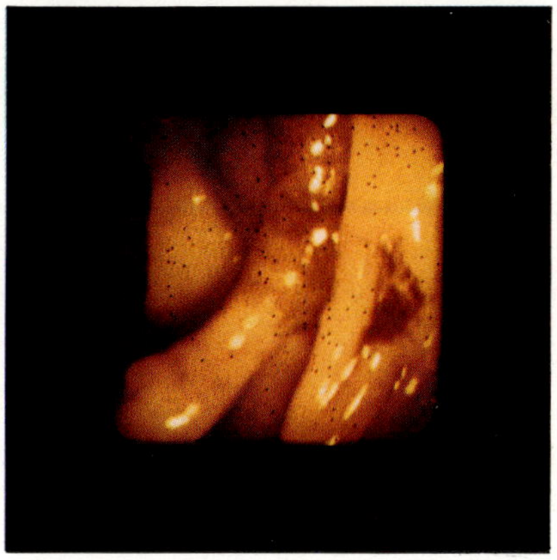

g. 287.
ssile polypoid growth the size
a walnut, with an uneven surface
vered with wine-red mucosa, in
sigmoid colon. On biopsy the
owth is soft, friable and
eds easily.

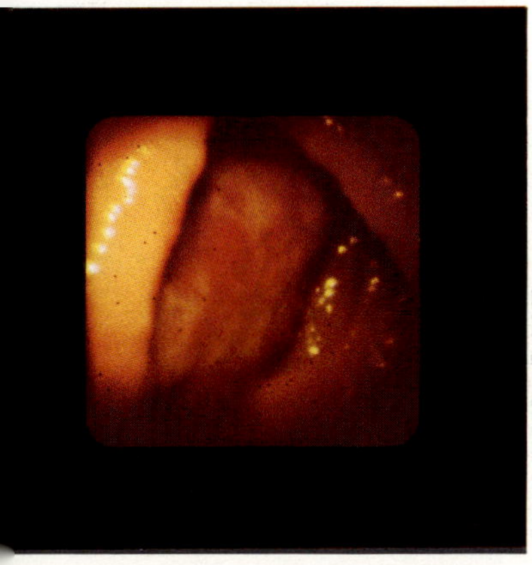

Fig. 288.
Coloscopic biopsy. Histologic diagnosis: papillary adenomatous polyp.

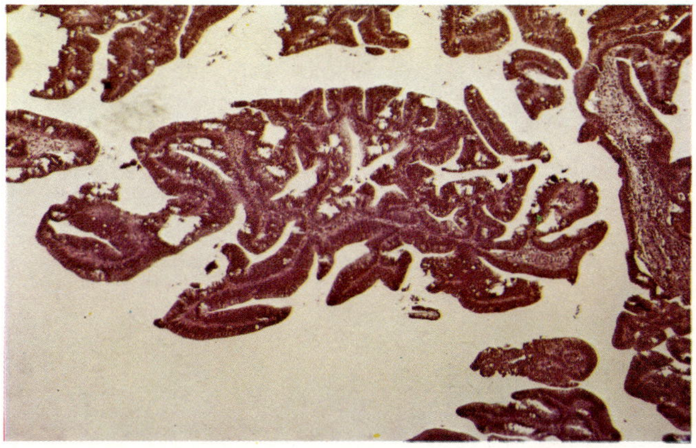

Fig. 289.
Barium enema. Single polyp of the colon. Round filling defect in the upper portion of the descending colon. The nearby bowel wall is normally contractile.

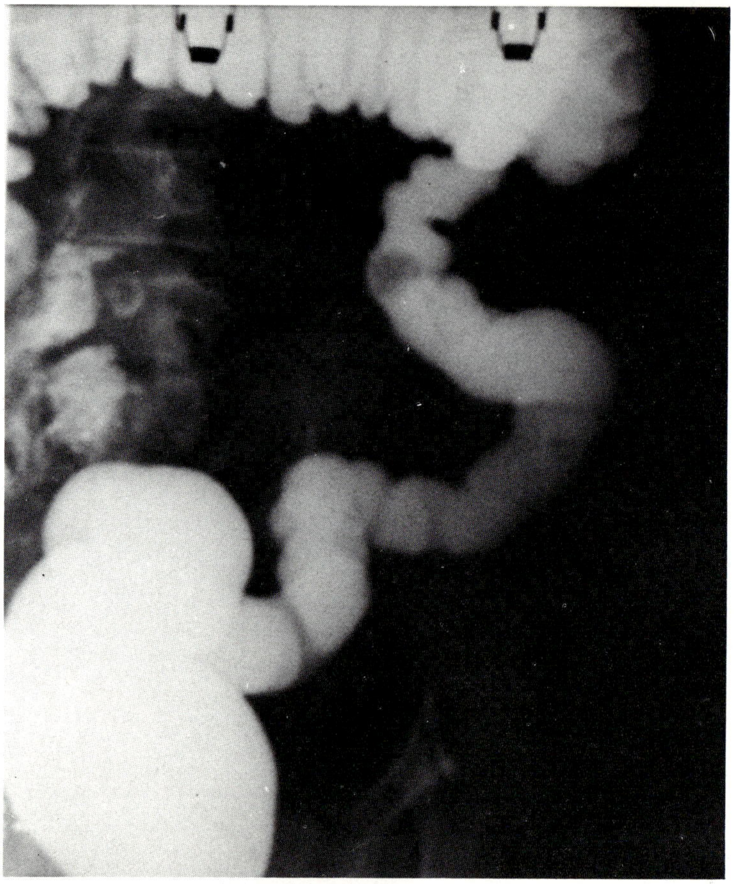

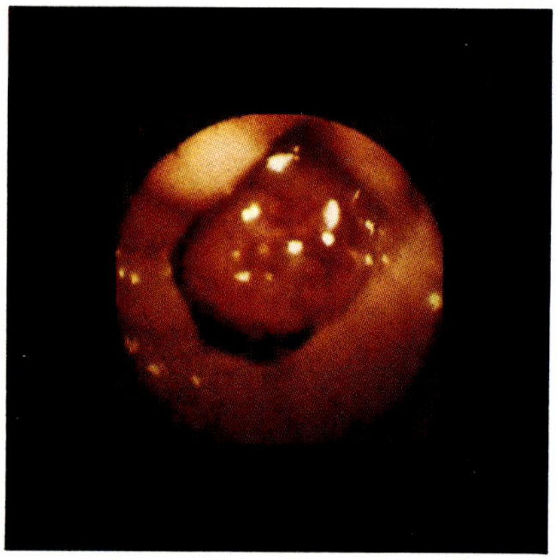

Fig. 290.
The peristaltic contractions of the bowel push forward the polyp. The polyp is the size of a walnut and has an irregular surface covered with wine-red, ulcerated mucosa which bleeds readily when touched by the fibrescope.

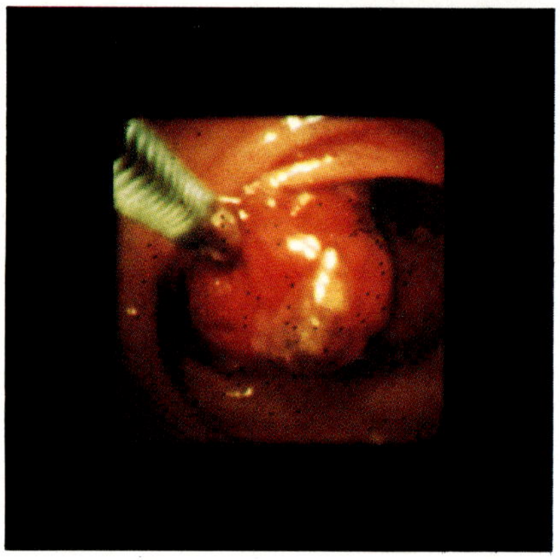

Fig. 291.
Manipulation of the polyp with the aid of the biopsy forceps displaces the bowel wall at its base thus showing it to be a sessile growth.

Fig. 292.
Coloscopic biopsy. Histologic diagnosis: adenomatous polyp with atypical glandular structures.

Fig. 293.
Detail of an atypical area at greater magnification.

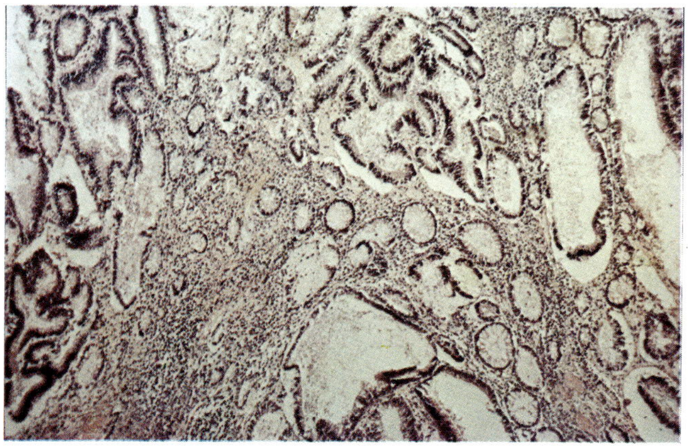

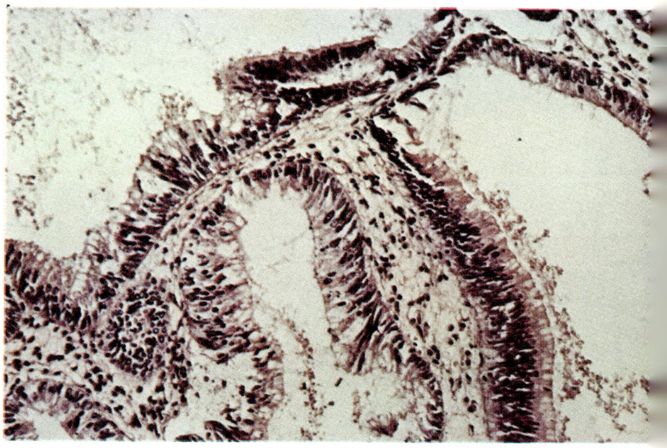

g. 294.
ɜeudolobulated sessile polypoid
owth at the sigmoid-descending
olon junction. The uneven surface
covered with pinkish mucosa
oon which there is a layer
mucus and fibrin. On palpation
ith the biopsy forceps the
olyp is soft and friable and is
sily penetrated by the jaws
the forceps. The endoscopic
cture is uncertain; the irregular
rface and brittleness of the
sion suggest a malignant
owth.

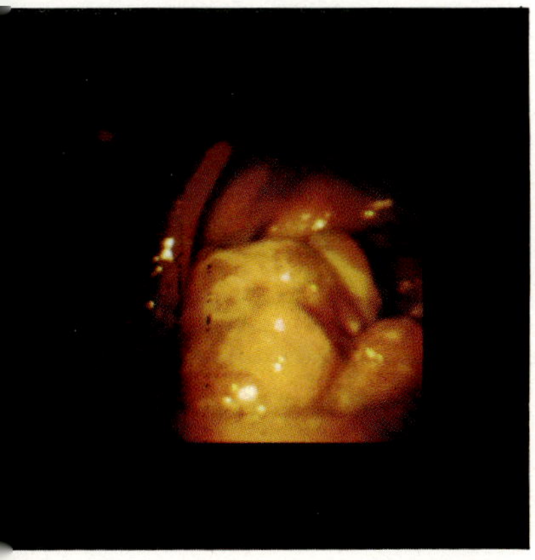

Fig. 295.
Coloscopic biopsy. Histologic diagnosis: adenomatous polyp with carcinoma in situ.

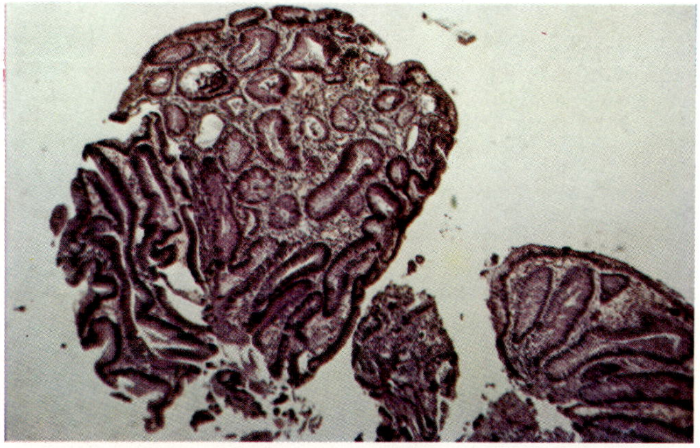

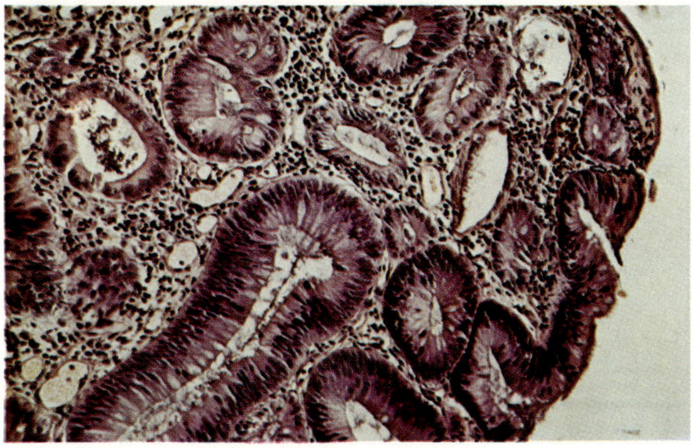

Fig. 296.
Same case. Detail of the carcinoma.

multiple polyps

Fig. 297.
Double contrast barium enema. Roundish, regular, sharply-outlined growths grouped together in the iliac-pelvic portion of the colon. The adjacent bowel wall is normally distensible.

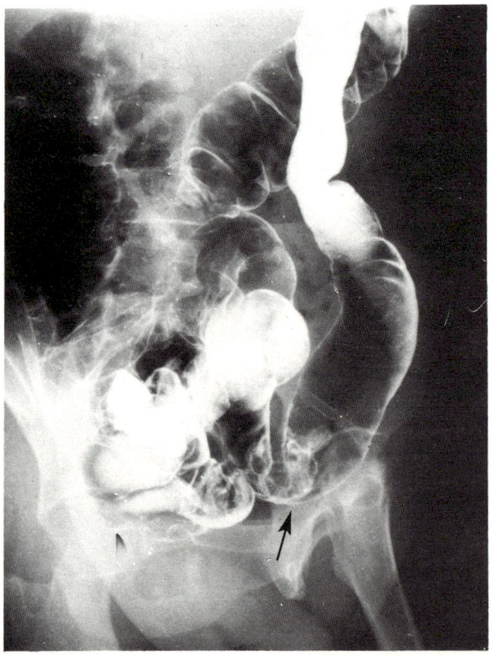

Fig. 298.
Adenomatous polyp. Hard, sessile polypoid growth with an irregularly ulcerated surface, the size of a hazel nut, in the sigmoid colon.

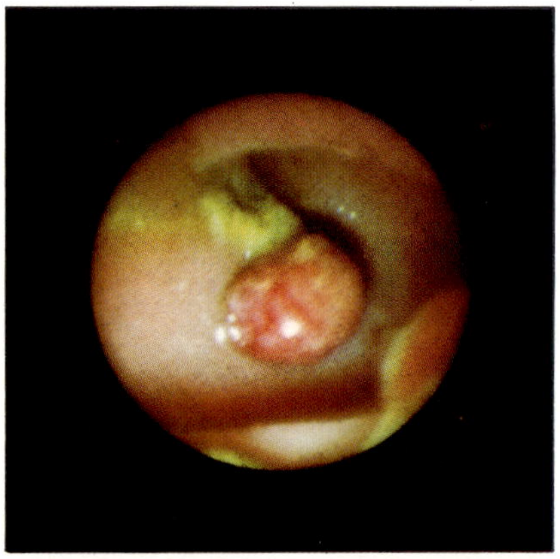

Fig. 299.
Same case. Another hard, pseudolobulated, short-stalked polypoid growth covered with greyish-pink, smooth, shiny muco in the descending portion of the sigmoid colon.

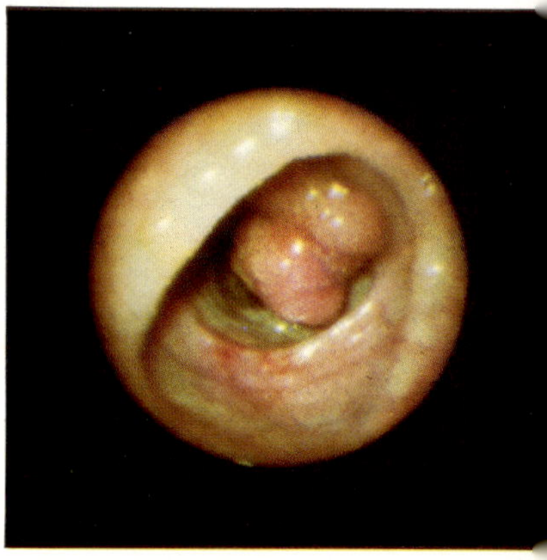

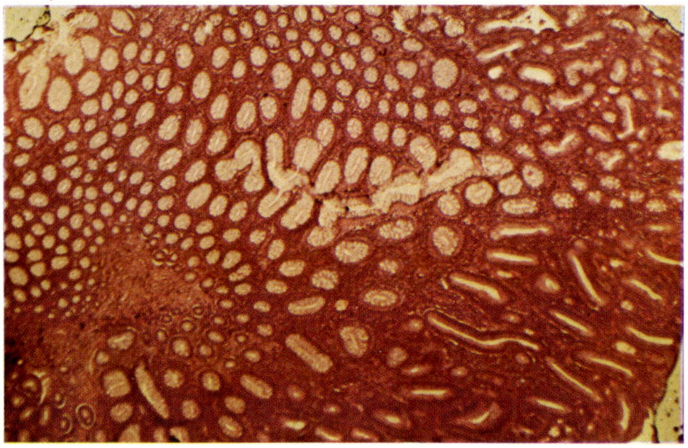

Fig. 300.
Coloscopic biopsy. Histologic diagnosis: adenomatous polyp.

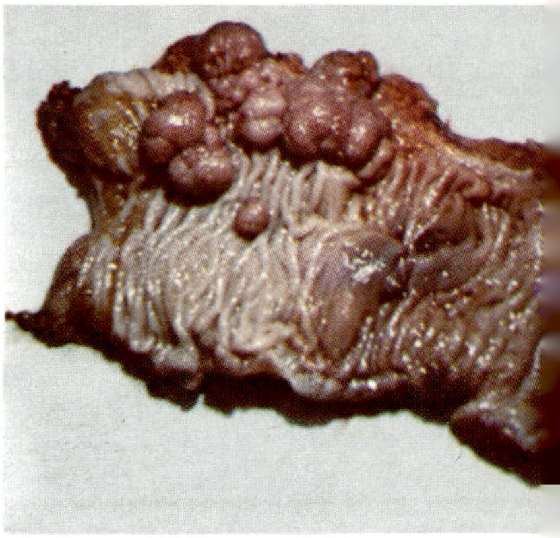

Fig. 301.
Surgical specimen. Piece of large intestine. Numerous sessile polyps of different shapes and sizes on the mucosa. The mucosa around the growths is normal. The bowel wall of the colon is normal. Histologic examination confirmed the endoscopic diagnosis of multiple adenomatous polyps.

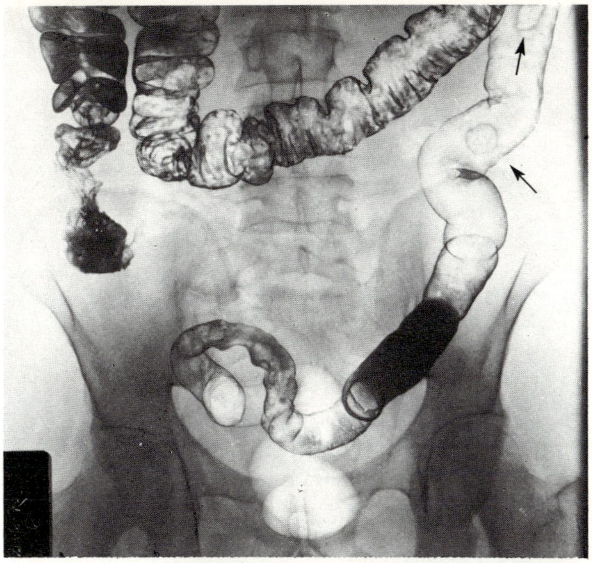

Fig. 302.
Double constrast barium enema. Two roundish growths of polypoid aspect, with regular, sharply-outlined contours, in the proximal and middle portions of the descending colon.

Figs. 303 & 304.
Endoscopic view of the same case. The two polyps are broad-based, the size of a hazel nut, covered with scarlet-red, finely irregular mucosa, and are hard and elastic on palpation with the biopsy forceps.

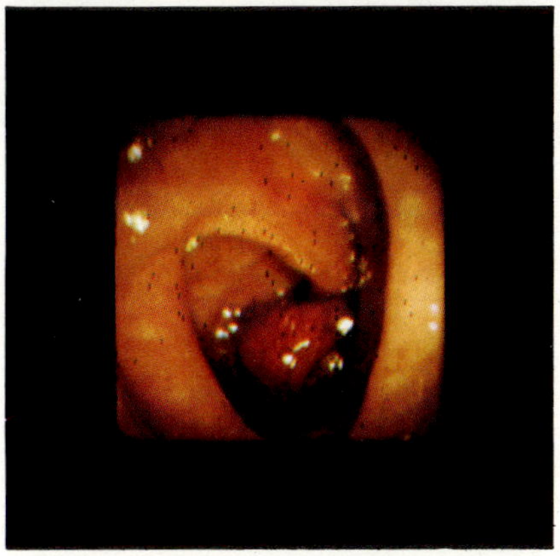

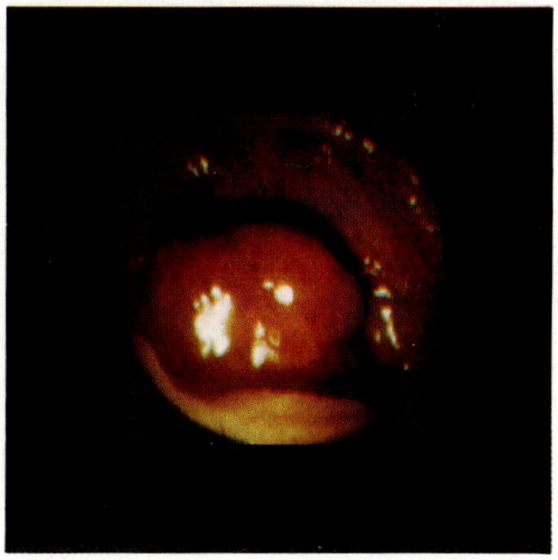

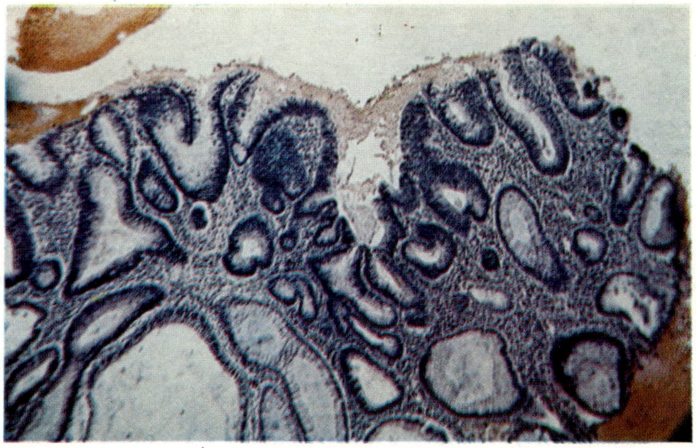

Fig. 305.
Coloscopic biopsy. Histologic diagnosis: adenomatous polyp. Note the sharp contrast between the normal mucosa and the mucosa with markedly atypical cellular features. Detail of the atypical area.

diffuse polyposis

Fig. 306.

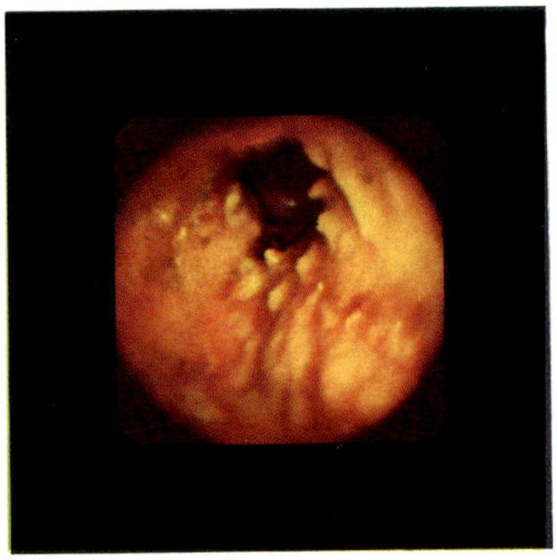

Fig. 307.

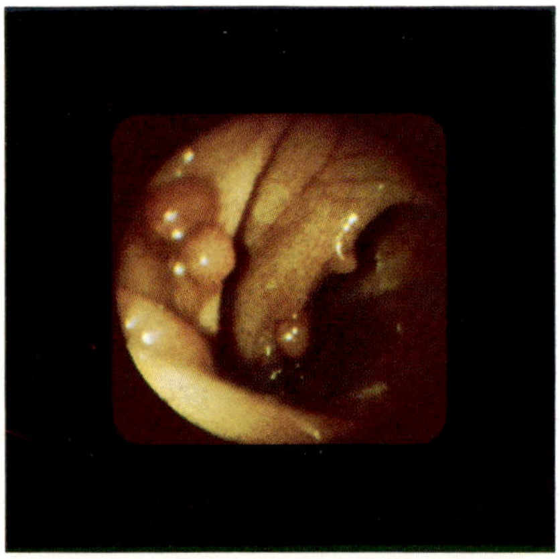

Fig. 308.

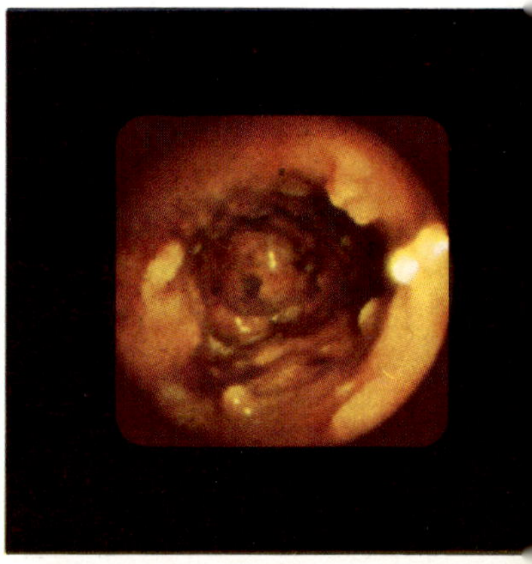

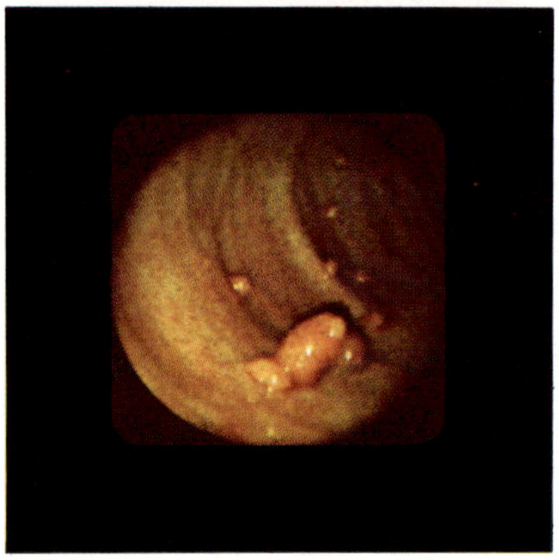

Fig. 310.

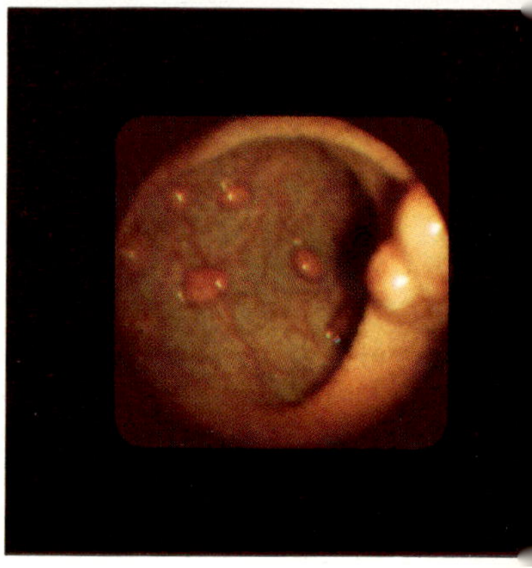

Fig. 311.

ig. 309.

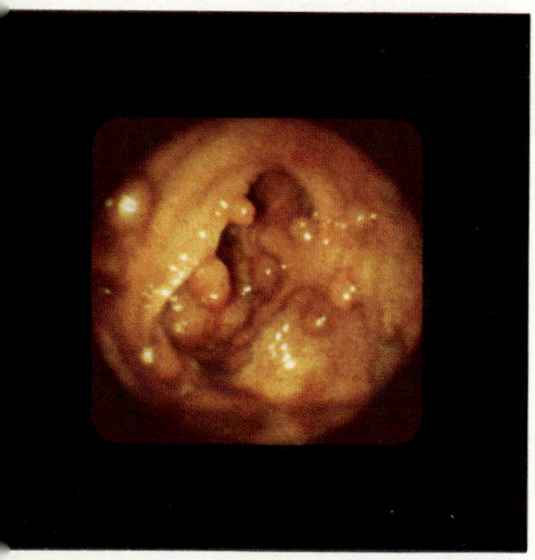

Numerous sessile polyps of different shapes and sizes in the sigmoid colon (Figs. 306 & 307), descending colon (Figs. 308 & 309). transverse colon (Fig. 310) and at the hepatic flexure (Fig. 311). Large, brittle, sessile growth with an uneven surface covered with ulcerated mucosa, in the right colon (Figs. 312 & 313). The bowel wall is everywhere normally distensible on insufflation. Histologic examination of the surgical specimen revealed diffuse polyposis of the large bowel and adenocarcinoma of the right colon.

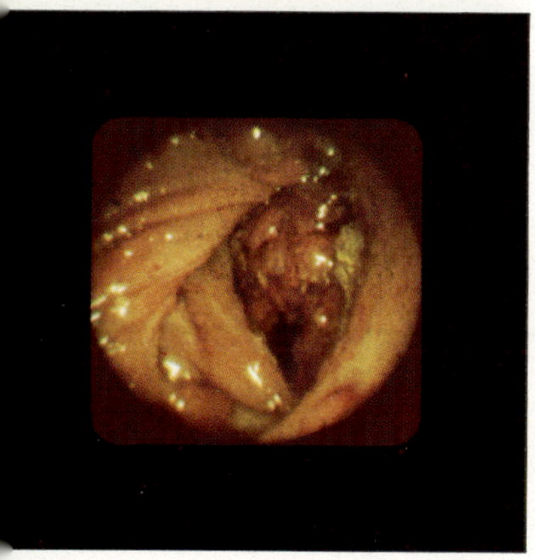

g. 312.

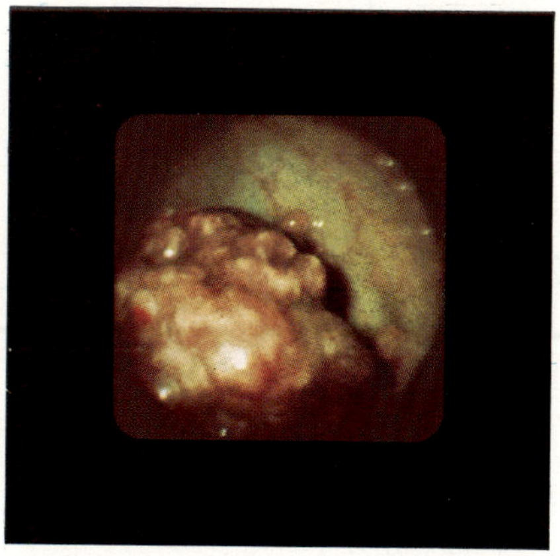

Fig. 313.

diffuse familial polyposis

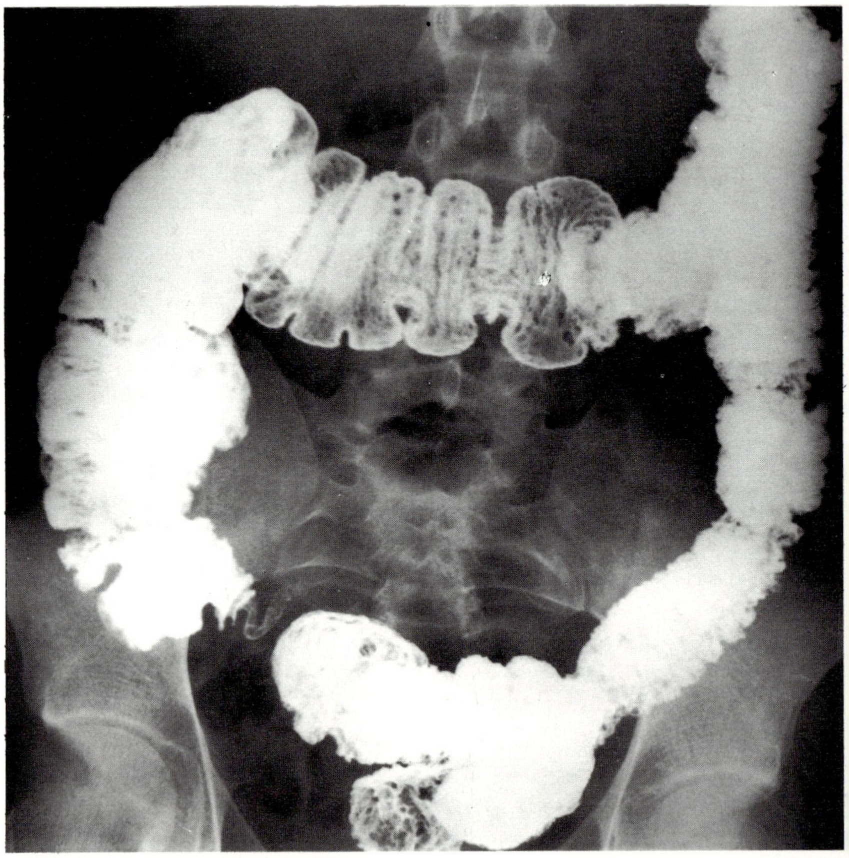

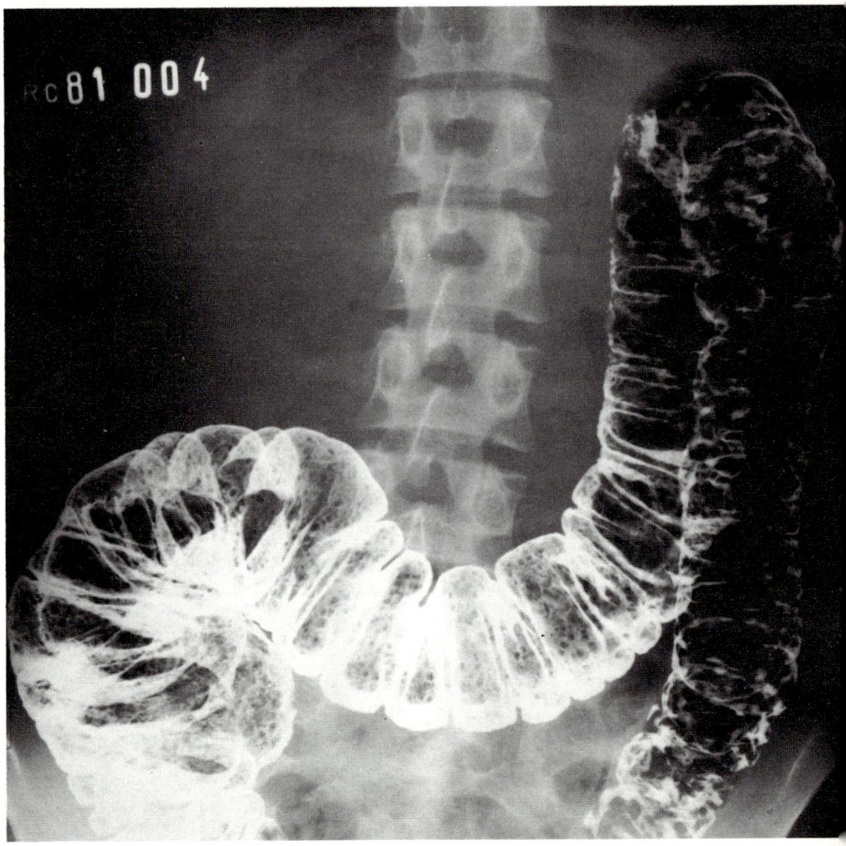

Double contrast barium enema:

Fig. 314.
large volume

Fig. 315.
after insufflation.

Innumerable round-shaped, regular and sharply-outlined filling defects carpet the whole large intestine.

g. 316.
ndoscopic view. Numerous small
ssile polyps on the wall of the
gmoid colon. By forced
sufflation of air into the lumen
the colon it is possible to
sualize the areas which are
affected by polyposis.

Fig. 317.
Same case. Endoscopic view of the sigmoid-descending colon junction.

Fig. 318.
Same case. Innumerable polypoid growths of different shapes are also present in the descending colon.

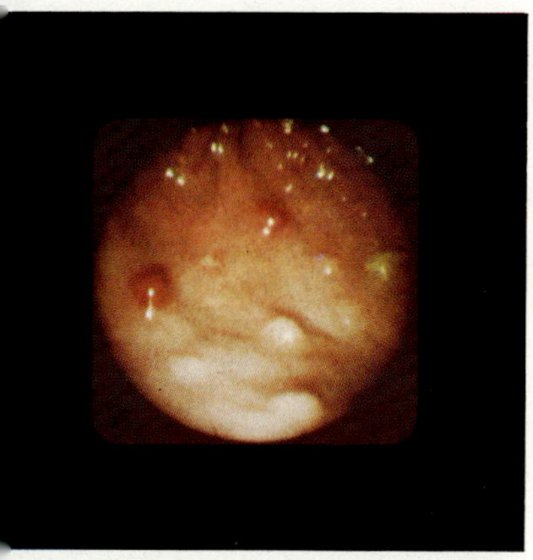

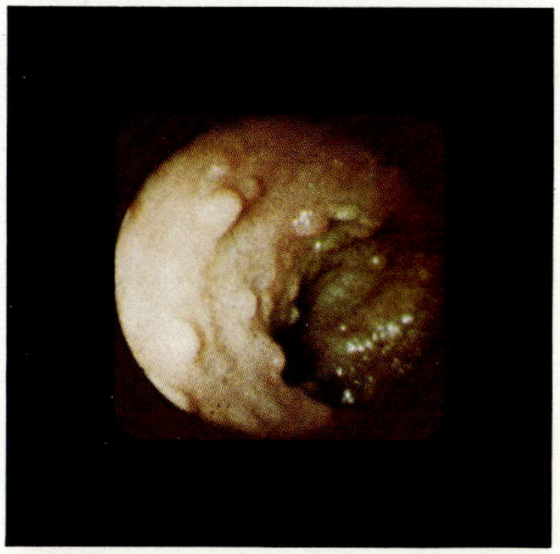

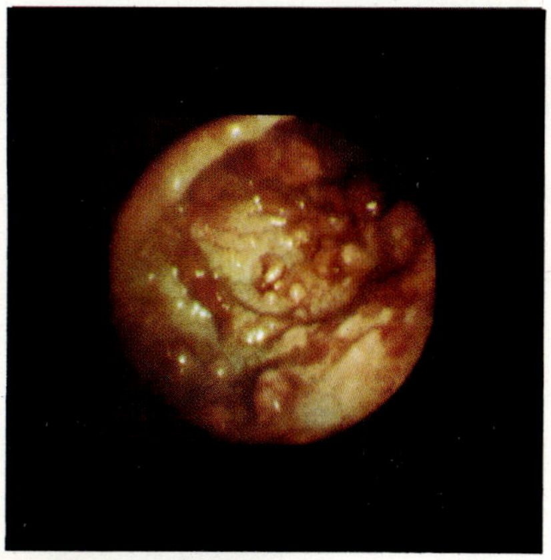

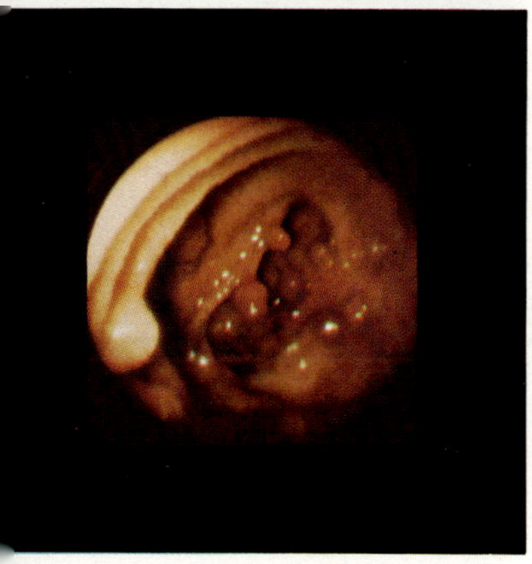

g. 319.
me case. View of the descending
lon at the left flexure. The
wel wall is distensible. The
ustral folds are preserved.

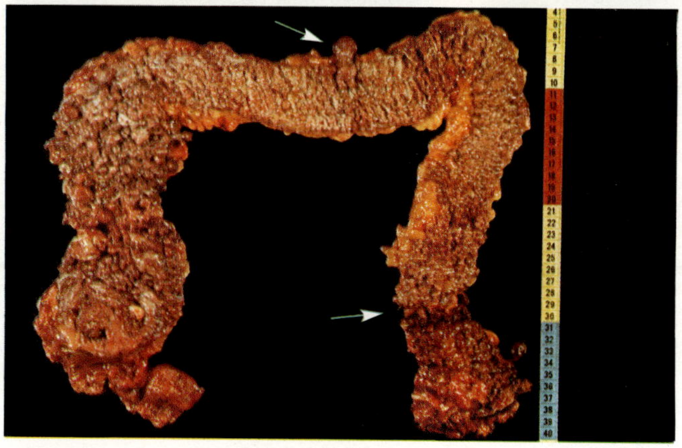

Fig. 320.
Surgical specimen. Total colectomy. Innumerable sessile polyps line the mucosa from the cecum to the rectum. Histologic examination showed some of the polyps to be adenocarcinomas (see arrows).

Fig. 321.
Double contrast barium enema showing innumerable regular, sharply-outlined filling defects, which carpet the colon. (Detail of the iliac-pelvic colon.)

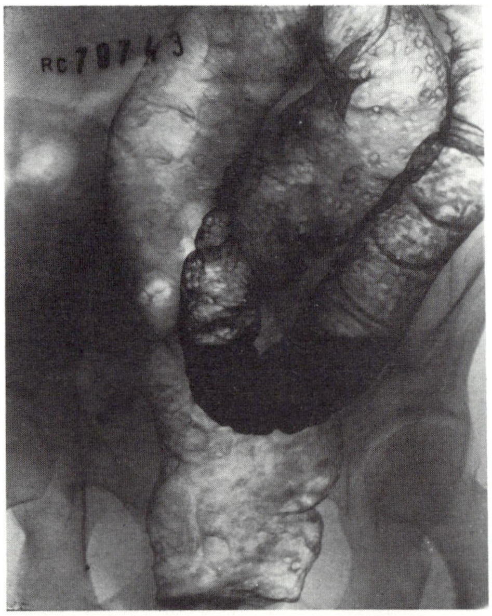

Fig. 322.
Endoscopic view. Numerous sessile polypoid growths of different shapes and sizes (2 mm to 1 cm) covered with smooth, shiny, greyish-pink mucosa on the wall of the sigmoid colon. Progression of the fibrescope is difficult on account of the reduced distensibility of the bowel wall even under high insufflation pressures, probably due to infiltration of the bowel wall.

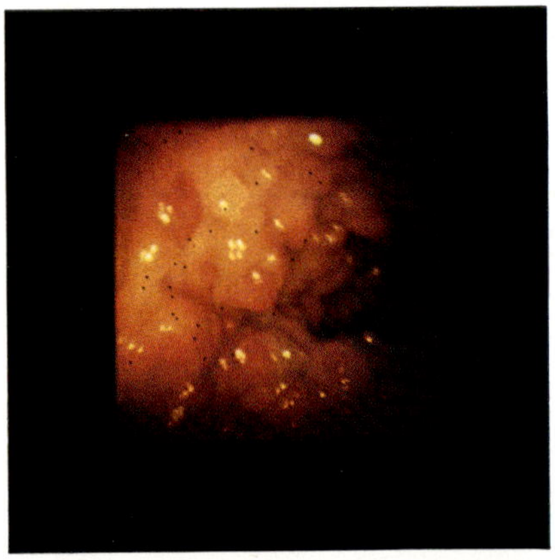

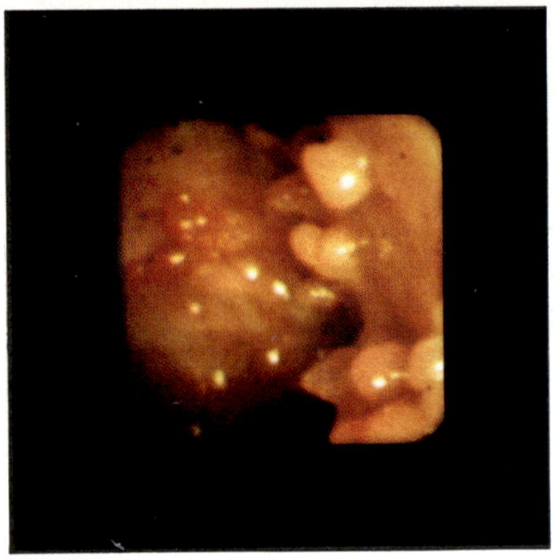

Fig. 323.
Same case. Endoscopic view of the descending colon.

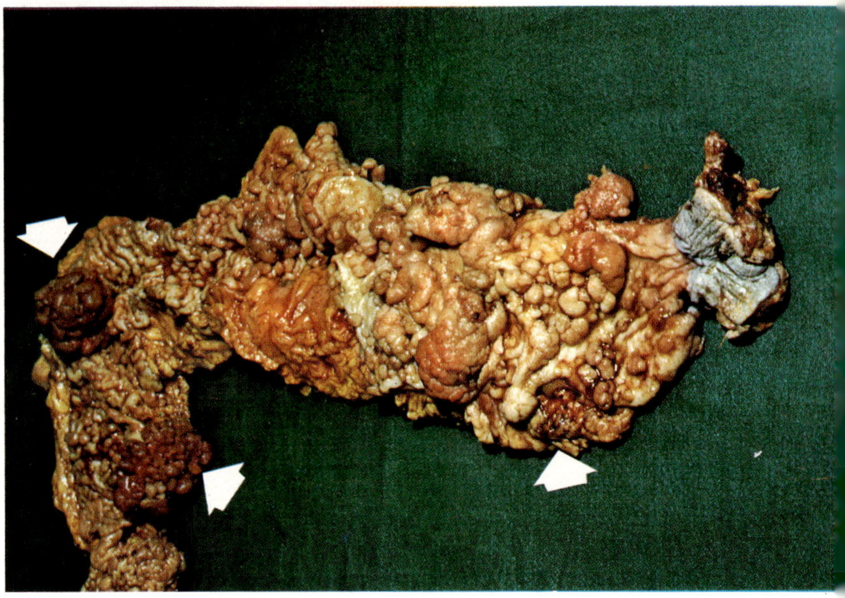

Fig. 324.
Surgical specimen. Portion of large intestine comprising the descending colon, the sigmoid colon and the rectum. Innumerable closely-packed sessile polyps of different shapes and sizes on the mucosa. In three places (see arrows) large cauliflower-like growths of greyish-red colour are present. On histologic examination these proved to be adenocarcinomas. The bowel wall close to the tumours is infiltrated.

uvenile polyposis

ıvenile polyps show the following endoscopic features:
- sessile or short-stalked;
- smooth-surfaced;
- friable (they separate easily from the bowel wall during endoscopic biopsy and as soon as the snare is tightened during endoscopic polypectomy);
- often associated with other new growths (adenomatous polyps, angiomas);
- mostly multiple, scattered or bunched together.

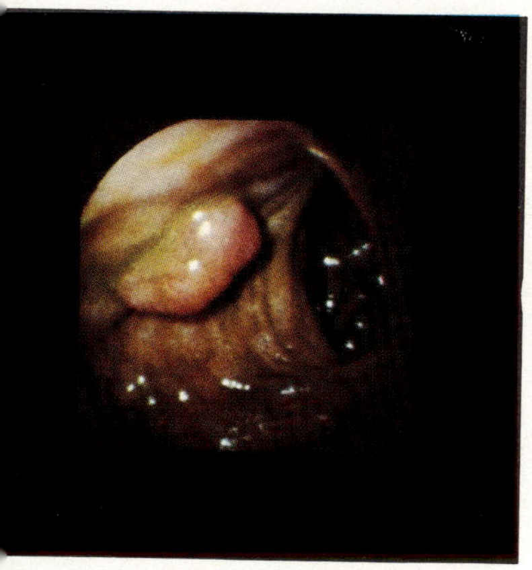

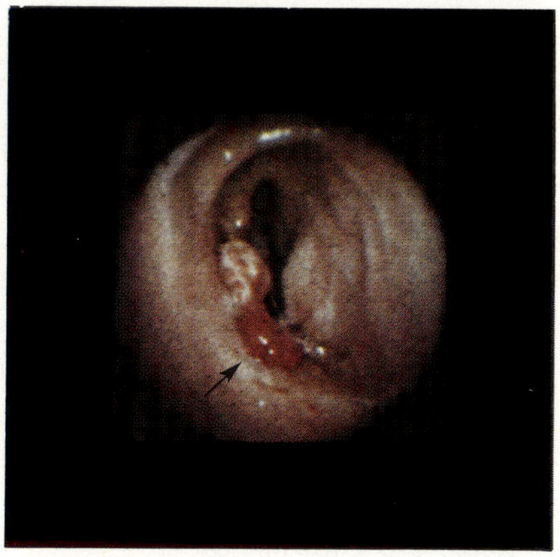

Fig. 326.
Juvenile polyp and angioma. Two sessile growths can be seen side by side in the transverse colon. One is smooth and greyish-pink in colour. The other is small, bright red, and bleeds easily (see arrow). A biopsy taken from the latter showed it to be an angioma. Bleeding after biopsy was severe and required endoscopic diathermocoagulation treatment.

;. 325.
ɔad-based, friable, polyp on ; wall of the descending colon. surface is smooth and covered th greyish-pink mucosa. The wel wall is normally distensible. stologic diagnosis: martomatous polyp.

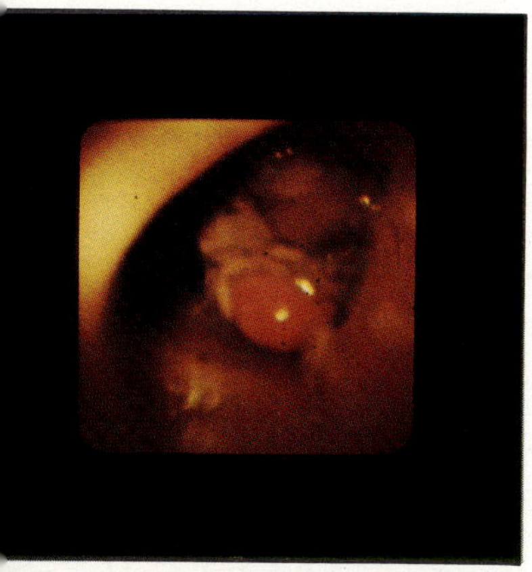

Fig. 327.
Multiple bunched polypoid growths. A number of the growths were removed bioptically and proved to be hamartoma-like polyps on histologic examination.

sentry polyp

Further radiologic and endoscopic investigations are mandatory whenever a new growth, no matter how small, is discovered at rectoscopy. One or more benignant polyps are often found below a higher-lying carcinoma. In many cases the benignant polyp is in a vanguard position with respect to the malignant lesion.

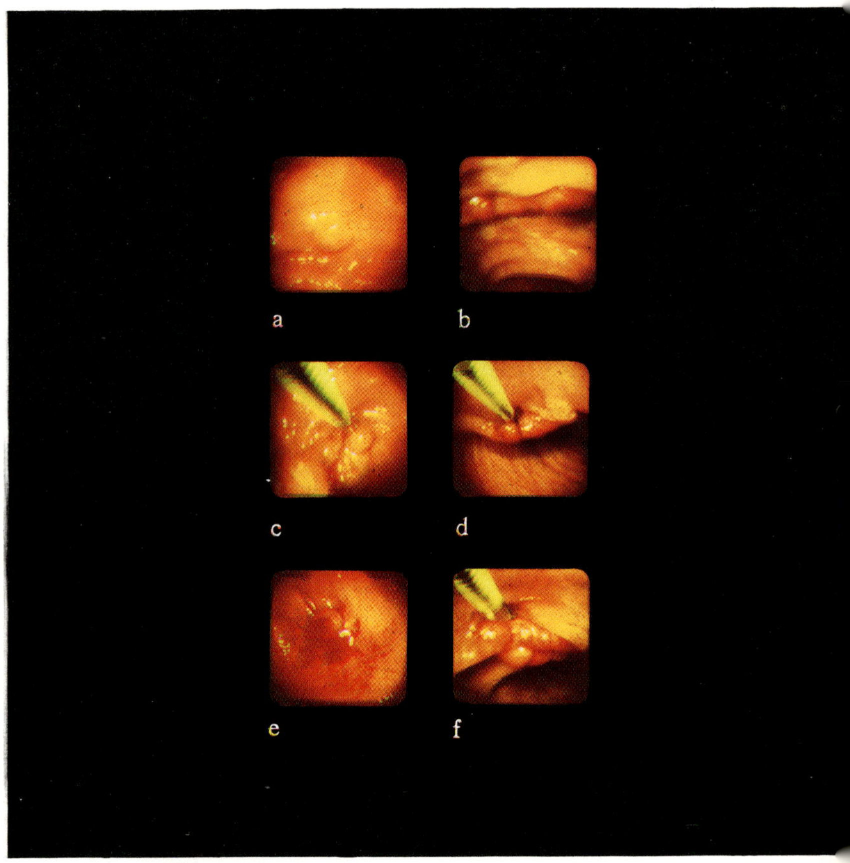

Fig. 328.
Serial photographs of a small sessile polyp of the sigmoid colon during and after biopsy. Histologic diagnosis: adenomatous polyp. A new growth of irregular appearance was found under a fold in the descending colon (Figs. d, e, f) and was visualized with the aid of the biopsy forceps. Histologic diagnosis: adenocarcicoma.

Fig. 329.
Surgical specimen. Portion of large intestine comprising the splenic flexure and the descending colon.
Histologic examination of the large cauliflower-like growth with central ulceration close to the splenic flexure showed it to be an adenocarcinoma. The small sessile polyp below the carcinoma in the descending colon proved to be an adenomatous polyp.

ig. 330.
mooth-surfaced sessile polyp
overed with greyish-pink
ıucosa.

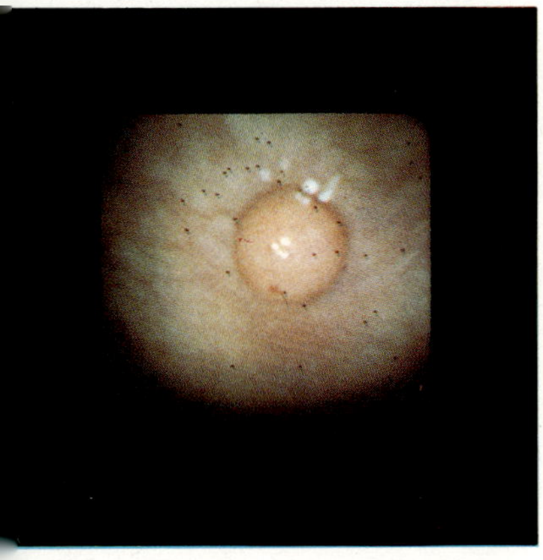

Fig. 331.
Same case. By pulling on the polyp with the biopsy forceps it can be shown that the mucosa around the base of the growth is normally elastic.
Endoscopic diagnosis: benignant polyp.

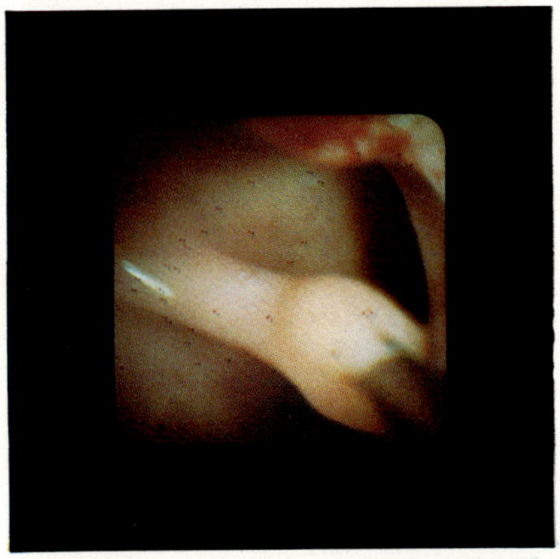

Fig. 332.
Same case. Another sessile growth of tough consistency can be seen close to the former. It has a finely uneven, ulcerated surface covered with mucus and fibrinous exudate. Its endoscopic appearance suggests a malignant growth.

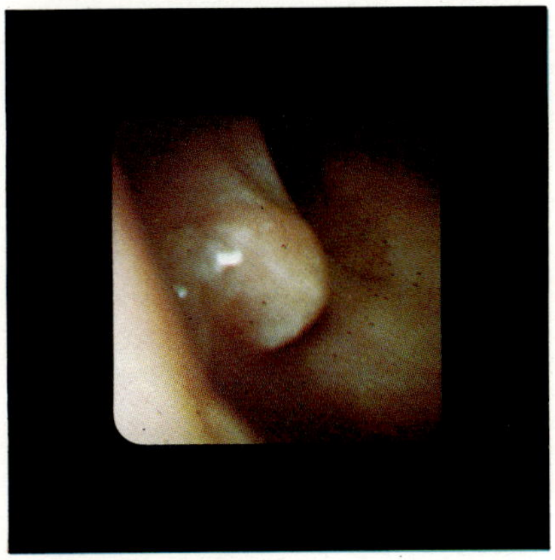

ig. 333.
ndoscopic view of the two growths
fter biopsy.

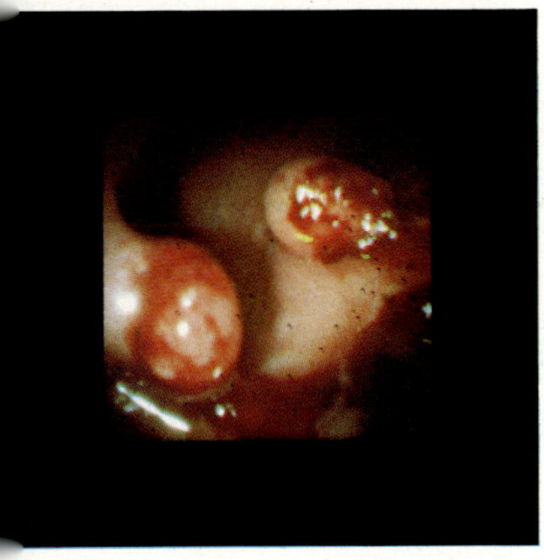

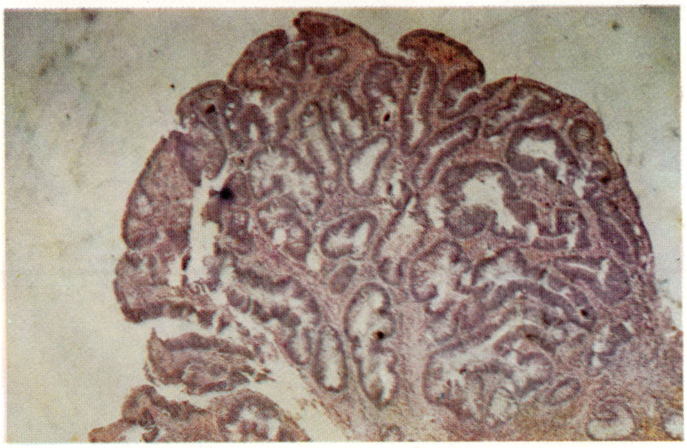

Fig. 334.

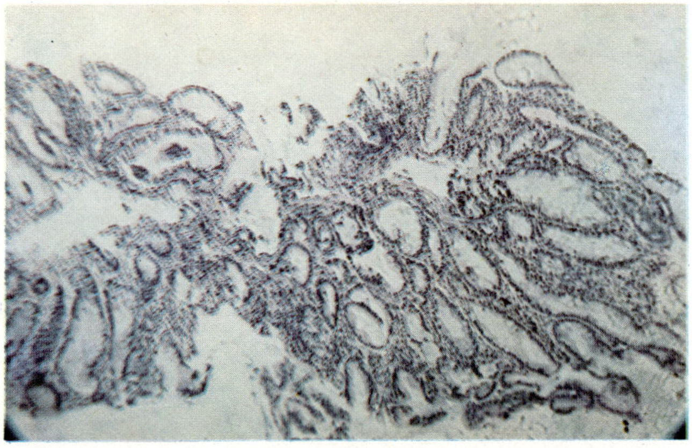

Fig. 335.

g. 334.
oloscopic biopsy of the polyp
Figs. 330 & 331. Histologic
agnosis: adenomatous
olyp.

g. 335.
oloscopic biopsy of the sessile
rowth in Fig. 332.
istologic diagnosis:
lenocarcinoma.

satellite polyp

polyp in diverticular disease

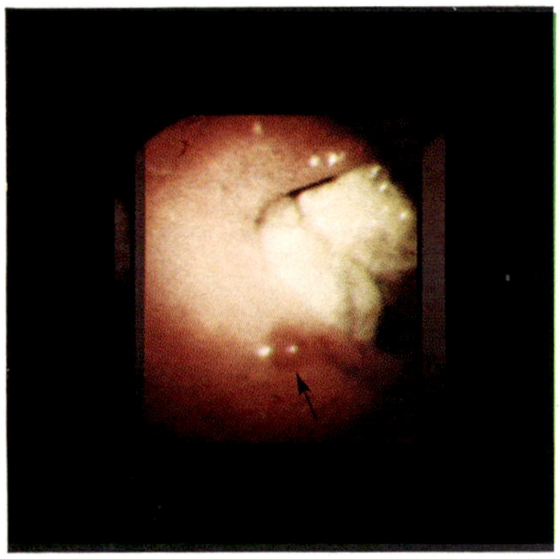

Fig. 336.
Malignant scirrhous new growth with an uneven surface of lardaceous aspect obstructing the lumen of the descending colon. On the summit of a mucosal fold there is also a small, hard, sessile polypoid growth the size of a millet seed, covered with smooth, shiny, pinkish-red mucosa. Histologic diagnosis: adenomatous polyp. (Courtesy of prof. A. Montori, Rome.)

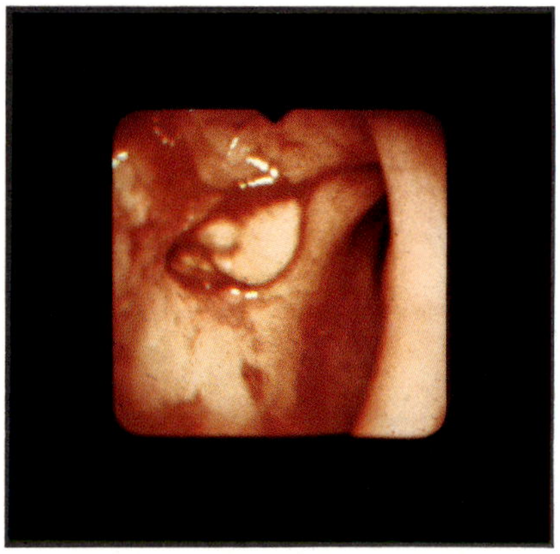

Fig. 337.
Diverticulosis of the colon.
Polyp in diverticulum. A diverticular cavity of the descending colon is filled almost completely by a smooth-surfaced sessile polpoid growth covered with greyish-pink mucosa. Histologic diagnosis: adenomatous polyp. (Courtesy of dr. J. R. Armengol-Miro, Barcellona.)

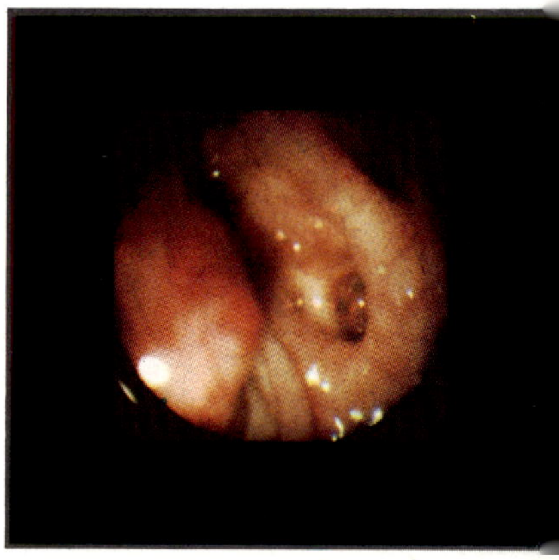

Fig. 338.
Adenomatous polyp with atypical features in a case of diverticulosis. Hard, elastic sessile polypoid growth with an uneven surface covered with hyperaemic mucosa in the distal portion of the sigmoid colon. The adjacent mucosa is also congested. Diverticula are also present.

Leiomyofibroma of the cecum

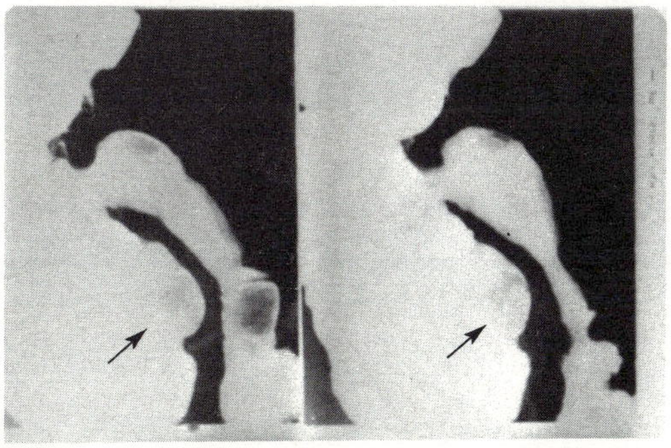

Fig. 339.
Barium enema. Small, round filling defect on the medial aspect of the cecum. The adjacent bowel wall is of regular aspect and normally elastic.

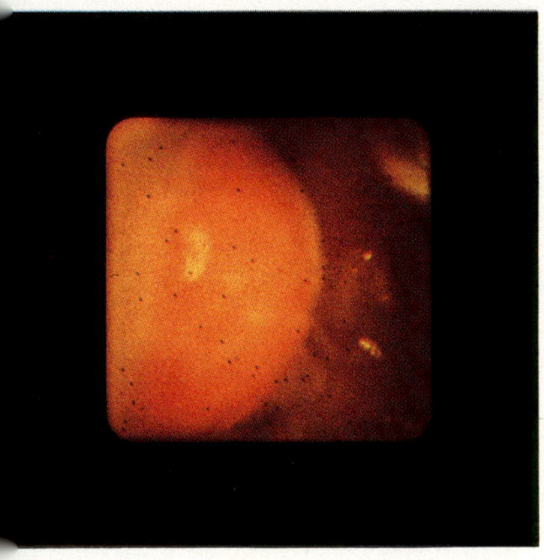

ig. 340.
lose-up, magnified endoscopic iew of the fundus of the cecum. lard, round, sessile growth the ze of a cherry, covered with nooth, shiny, pink mucosa. Aimed iopsy is difficult on account the obility and toughness of the rowth.

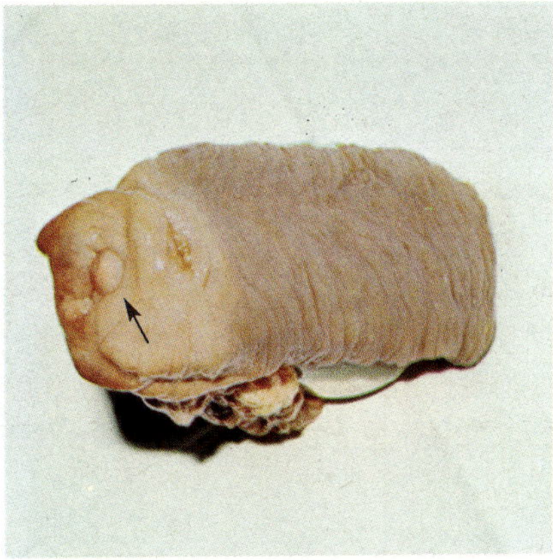

Fig. 341.
Surgical specimen, including the cecum and ascending colon. There is a hard, sessile, smooth surfaced growth 0.5 cm in size covered with normal mucosa in the cecum. Histologic diagnosis: leiomyofibroma of the cecum.

Polypoid plical hyperplasia

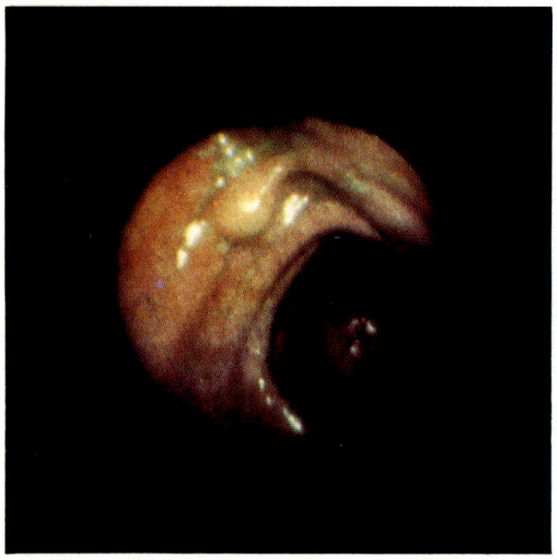

Fig. 342.
Elevation of polypoid aspect, about 5 mm in size, on the wall of the descending colon. It is covered with greyish-pink, smooth, shiny mucosa. On palpation with the biopsy forceps the lesion is elastic and mobile together with the adjacent mucosa. It flattens somewhat on insufflation of the bowel.

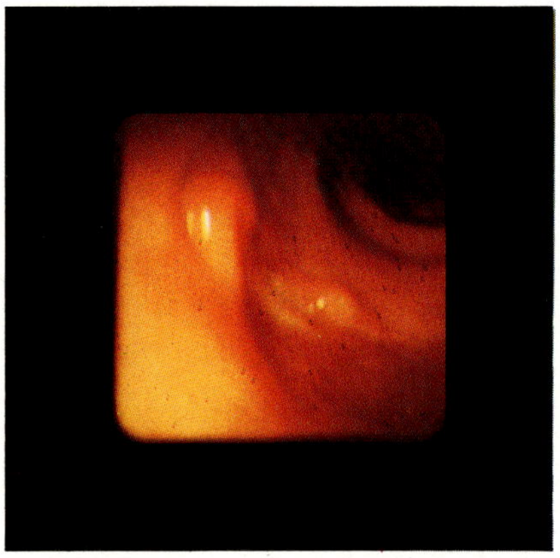

Fig. 343.
The mucosa of the descending colon forms elevations of polypoid aspect, covered with normal mucosa.

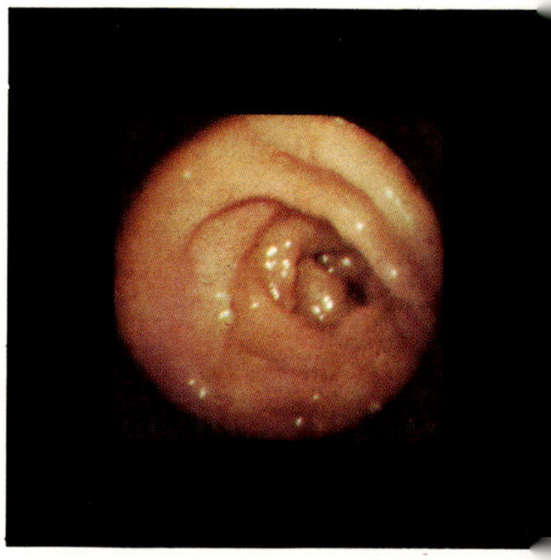

Fig. 344.
The mucosa of the descending colon forms medium-thick and thick, elastic folds covered with normal mucosa. Insufflation of air flattens these large folds only in part.

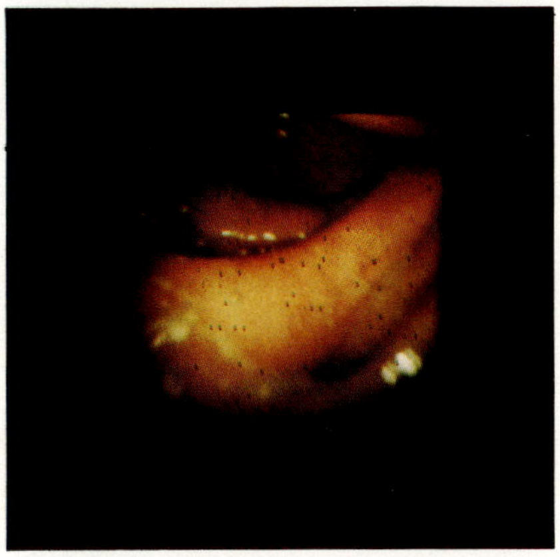

Fig. 345.
Large semilunar fold in the descending colon. The smoothness, pinkish colour and elasticity of the mucosa and the fact that air insufflation flattens the fold suggest plical hyperplasia. Histologic examination showed it to consist of mucosal tissue of the large bowel.

Angiomas

Fig. 346.
Angioma of the sigmoid colon. Small, bright red, soft elevation on the wall of the sigmoid colon.

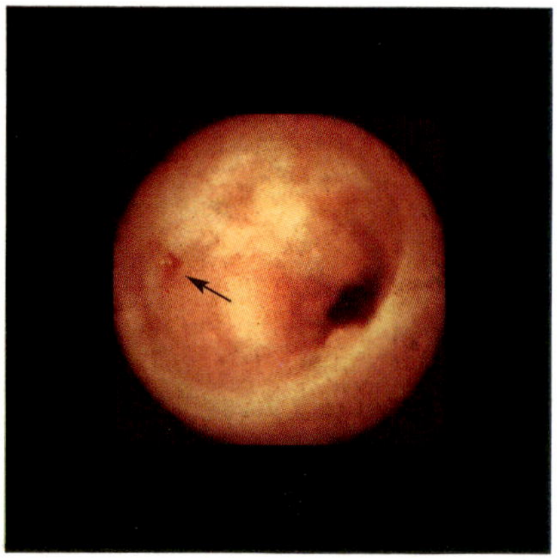

Fig. 347.
Cavernous angioma of the sigmoid colon. Semi-spherical, soft, scarlet-red excrescence covered with mucus in the distal portion of the sigmoid colon. Biopsy was followed by copious bleeding which could not be arrested by instilling epinephrine but only by electrocoagulation.

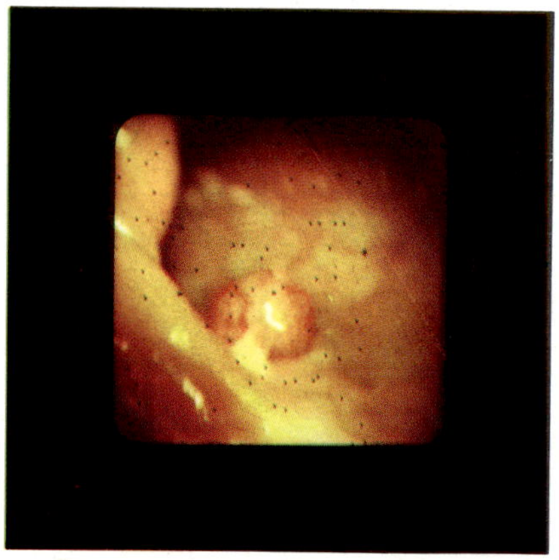

Fig. 348.
Same case. Site of the angioma after electrocoagulation. The excrescence has flattened completely.

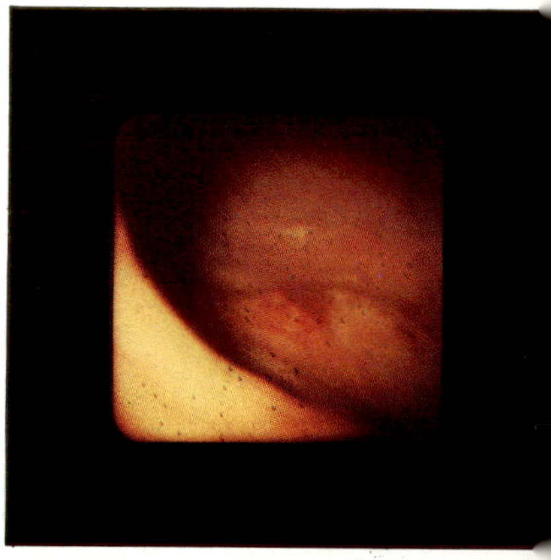

'ig. 349.
\ bright red patch which shows up
hrough the smooth shiny mucosa
s hidden under a circular
old close to the left flexure.
Histologic examination of the
urgical specimen revealed
aemangiomatosis of the splenic
lexure and transverse colon.

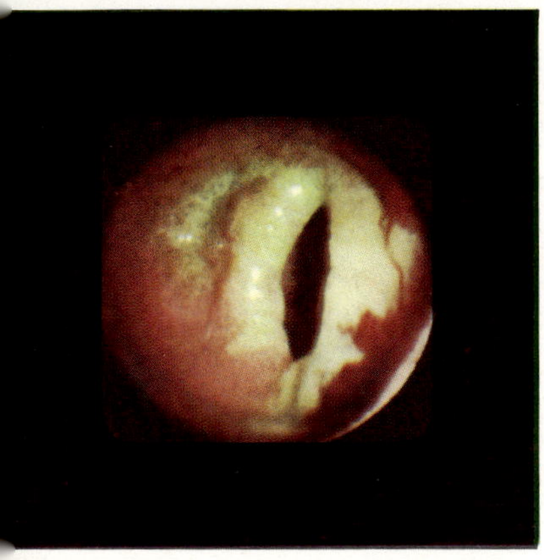

Fig. 350.
Large, soft, bright red angioma on a semilunar fold of the distal portion of the sigmoid colon.

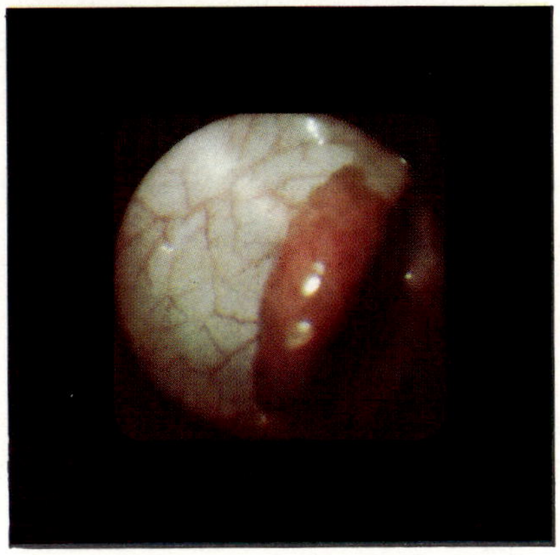

Varices of the large intestine

'ig. 351.
oft, protruding swelling in the
ımen of the descending colon.
'he bluish discolouration of the
ıucosa is due to the presence
f underlying varices. First grade
arices of the large intestine.

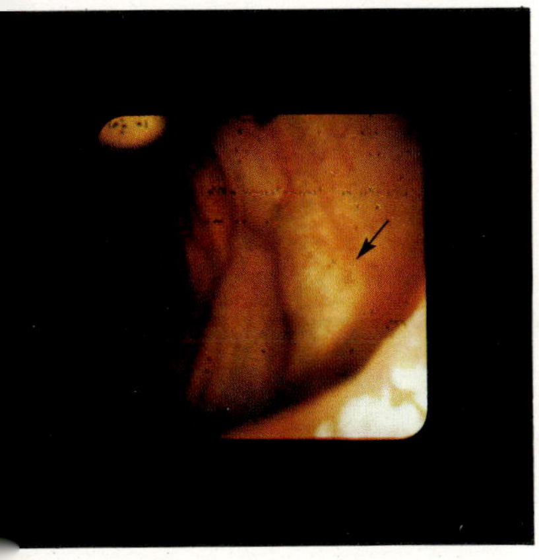

Fig. 352.
Turgid, tortuous varicosity, about
3 cm long, at the splenic
flexure. The surrounding mucosa
is pinkish, smooth and shiny.
Second grade varices.

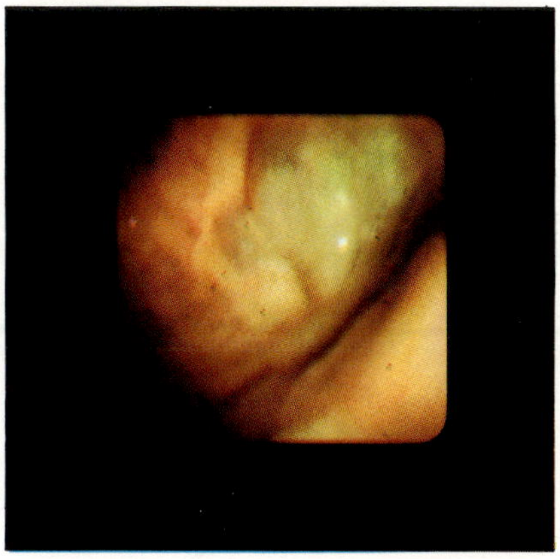

Fig. 353.
Sharply defined, purplish patch in
the middle portion of the
transverse colon. This endoscopic
picture is attributable to
venous stasis, probably due to
extrinsic compression.
This type of finding is common
in tumours of the cecum and
ascending colon.

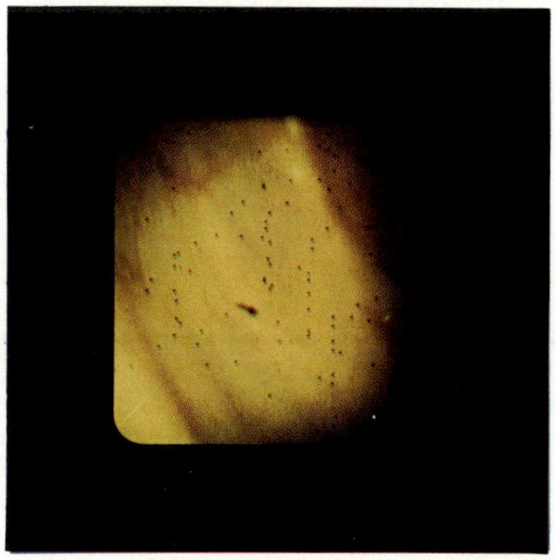

Carcinoma

Fig. 354.
Carcinoma of the descending colon (early stage). Sessile, pseudolobulated lesion covered with smooth greyish-pink mucosa on the wall of the descending colon which is also lined with smooth, greyish-pink mucosa. The mucosa around the base of the growth does not appear to be infiltrated.

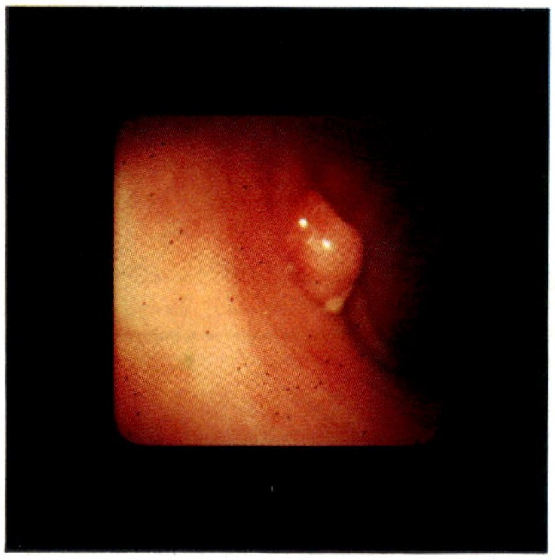

Fig. 355.
Coloscopic biopsy. Histologic diagnosis: adenocarcinoma.

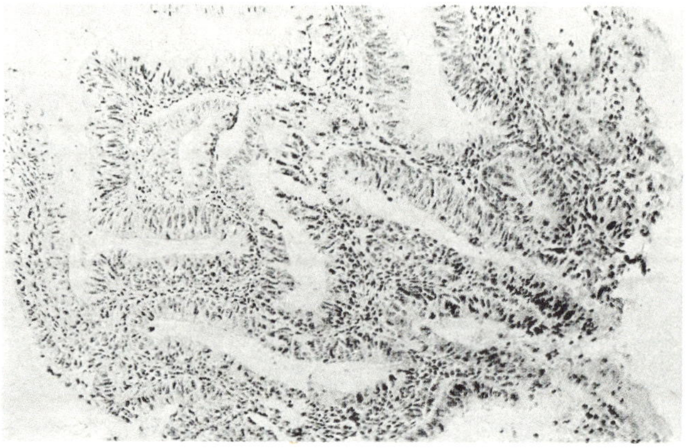

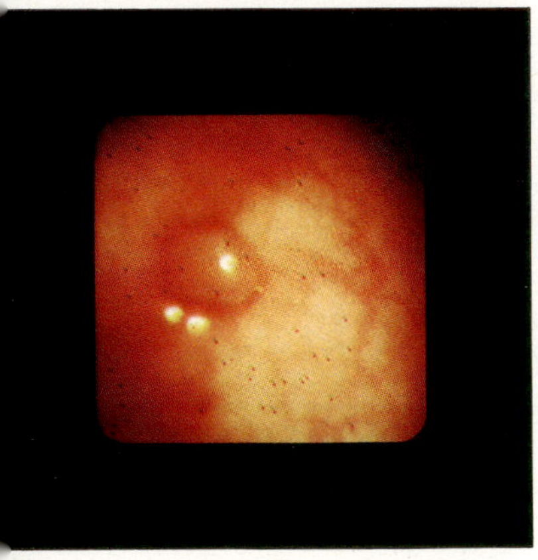

ıg. 356.
arly carcinoma of the sigmoid
ɔlon. Small sessile growth, 0.5 cm
ı diameter, covered with smooth,
eyish-pink mucosa. The endoscopic
spect of the lesion suggests an
lenomatous or hyperplastic sessile
ɔlyp (considering the patient's age).
anipulation with the biopsy forceps,
ɔwever, yielded other significant
ata. The growth proved to be of
ugh consistency, bled easily, and
as seen to infiltrate the nearby
ucosa. Aimed biopsy of the base
nd body of the lesion revealed a
on-infiltrating malignancy
denocarcinoma).

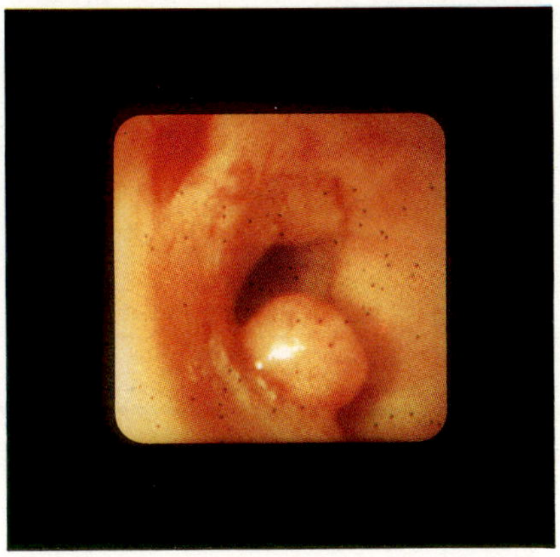

Fig. 357.
In this case the endoscopic findings suggest a malignant growth. The surface is finely irregular. The base is broad and appears to infiltrate the surrounding mucosa. The bowel wall is moderately stiff and is not easily distensible by insufflation. Histologic examination of biopsy specimens confirmed the endoscopic findings (adenocarcinoma).

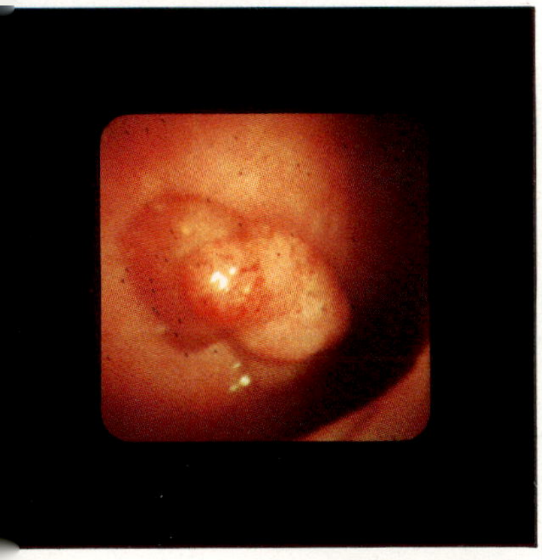

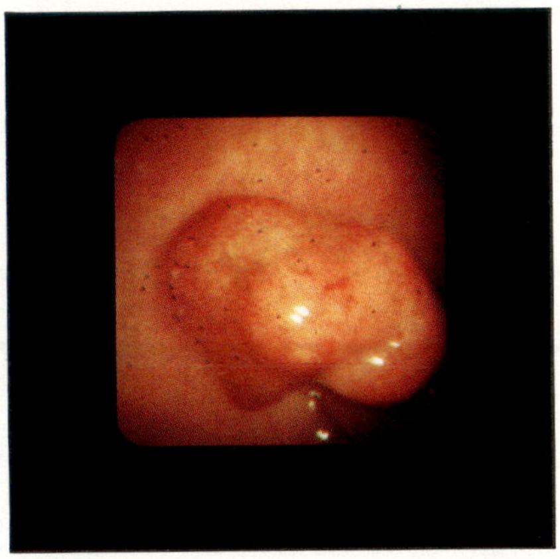

Fig. 358.
Broad-based, pseudolobulated new growth at the sigmoid-descending colon junction covered with pinkish, smooth, shiny mucosa.

Fig. 359.
Close-up view of the same case. Pseudolobulation and infiltration of the bowel wall are discernible.

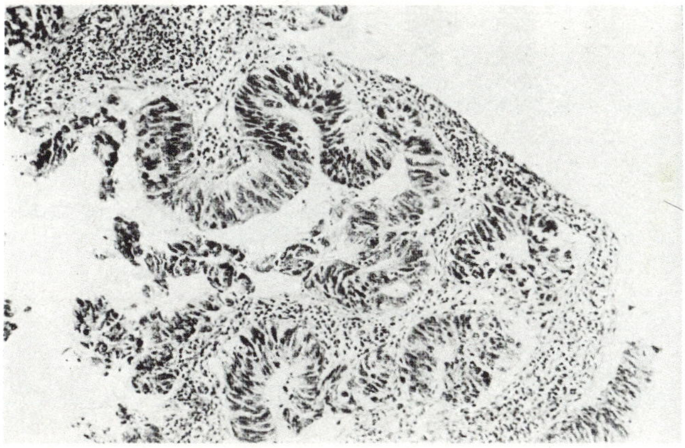

Fig. 360.
Coloscopic biopsy. Histologic diagnosis: adenocarcinoma.

Fig. 361.
Large fungoid growths in the descending colon. Ulcers with irregular, undermined borders are visible in the foreground.

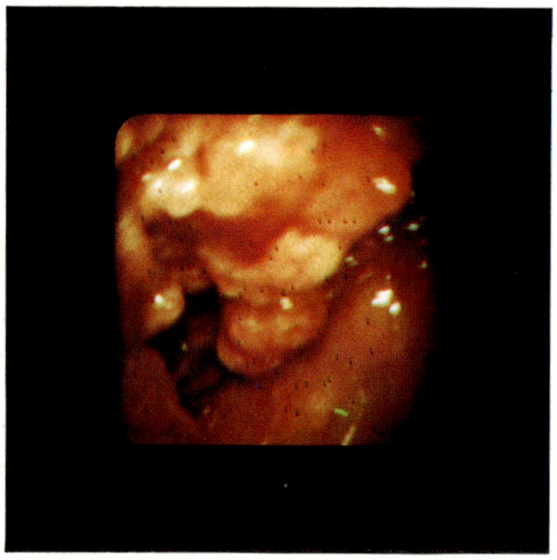

Fig. 362.
Ulcerated bleeding mammillated growths infiltrating and stenosing the large bowel at the sigmoid-descending colon junction.

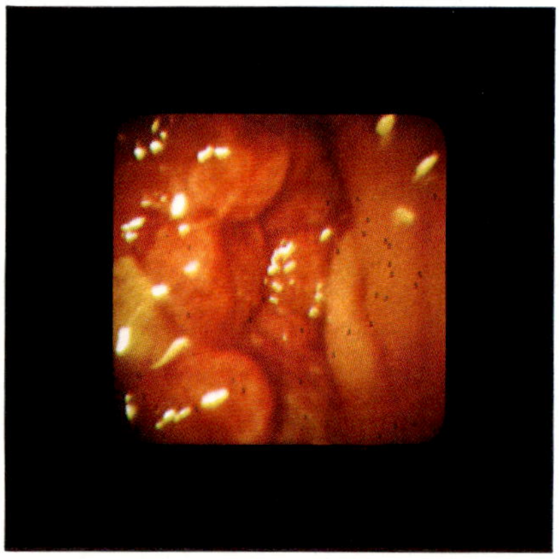

Fig. 363.
Large pseudolobulated fungoid growth occupying the entire lumen of the descending colon.

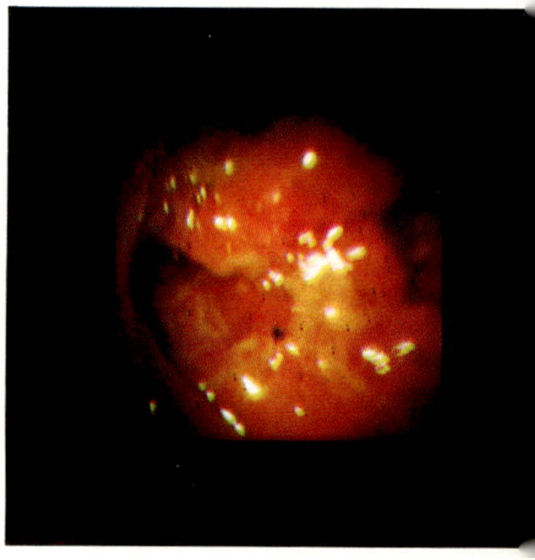

Fig. 364.
Stenosing fungoid carcinoma. Poutruded « napkin ring » lesion stenosing the bowel lumen at the sigmoid-descending colon junction. Shallow ulcers on the mucosa of the lower part of the tumour.

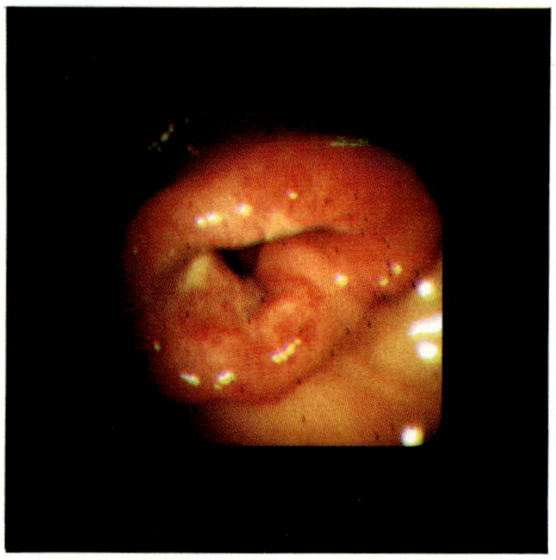

Fig. 365.
Same case. Bleeding after coloscopic biopsy.

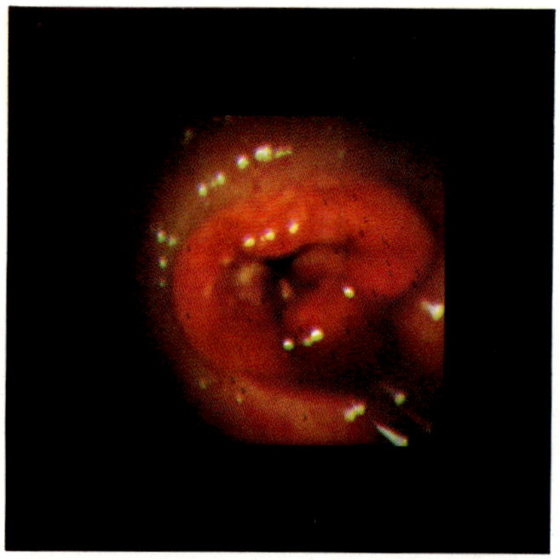

ig. 366.
nfiltrating tumour. Large, stiff, orseshoe-shaped fold covered with nely ulcerated mucosa infiltrating he middle third of the descending olon. The bowel lumen is patent.

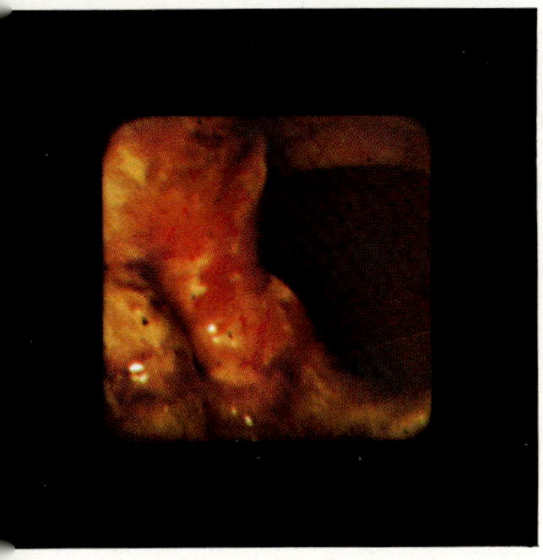

Fig. 367.
The wall of the middle third of the descending colon is rigid, fixed and involved by a tubular infiltrating heteroplastic lesion about 4 cm long.

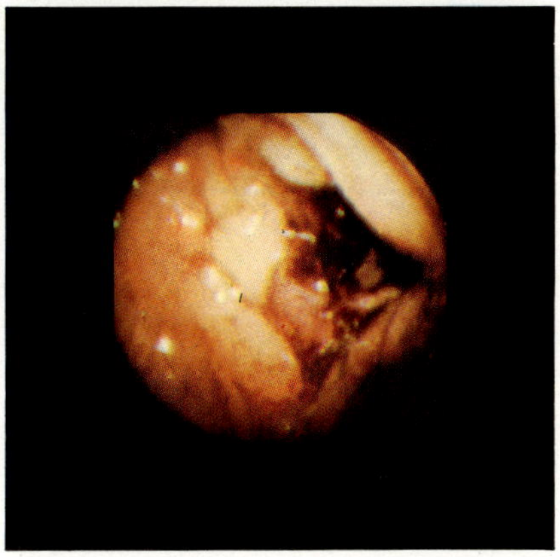

Fig. 368.
Broad-based fungoid growth which infiltrates the bowel wall at the hepatic flexure surrounded by blood-free faeces.

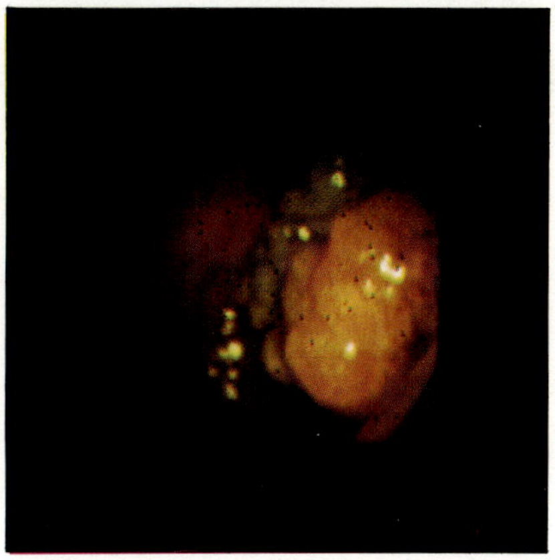

ig. 369.
ungoid tumour. Greyish, shiny, ard scirrhous growth protruding in ne lumen of the descending colon.

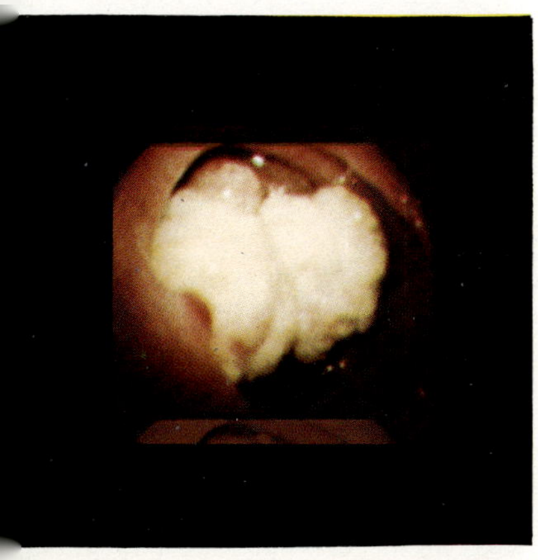

Fig. 370.
Fungoid tumor. Pseudolobulated lesion covered with pinkish, shiny mucosa in the lumen of the descending colon.

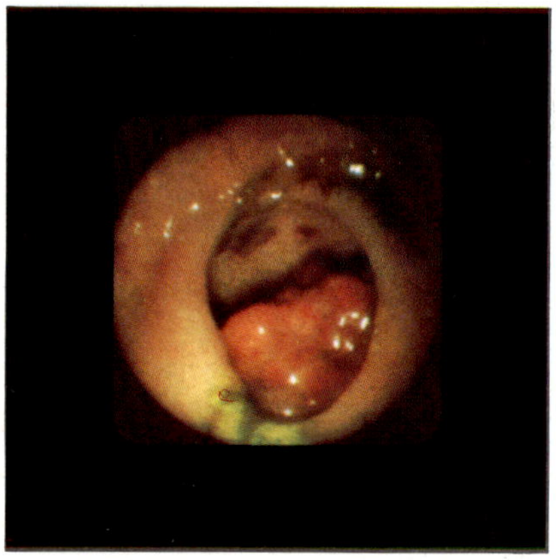

Fig. 371.
Same case. Base of the tumour visualized after distension of the bowel by insufflation. The first endoscopic view suggested a polyp rather than cancer.

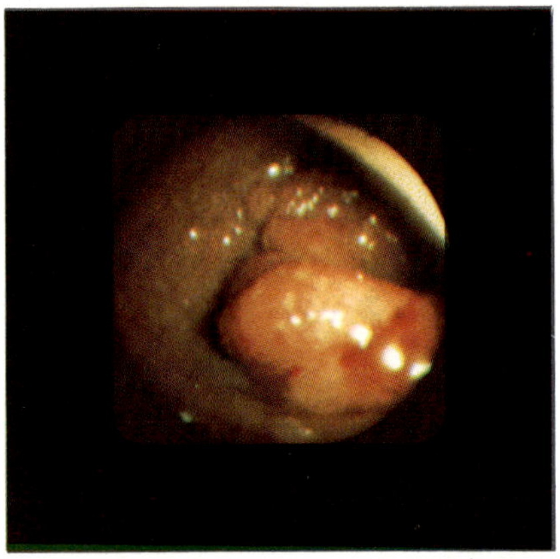

Fig. 372.
Fungoid tumour. Blackberry-like tumoural growth covered with ulcerated mucosa in the cecum.

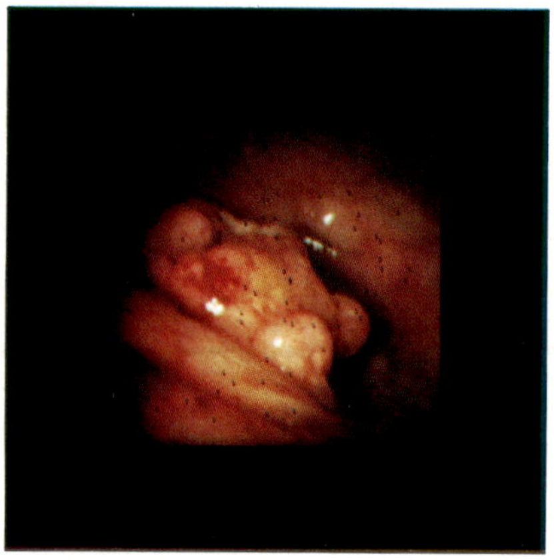

Fig. 373.
Close-up view of the same case. The biopsy forceps can be used not only to take biopsies but also to displace the lesion and thus evaluate its size, especially the width of its base, and the extension of infiltration.

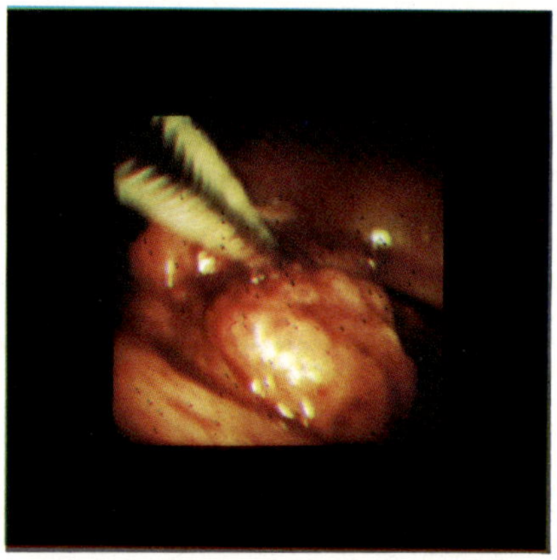

Fig. 374.
Fungoid tumour. Neoplastic growth occupying the sigmoid colon. The mucosa is ulcerated. The surface of the lesion is partly coated with muco-fibrinous exudate.

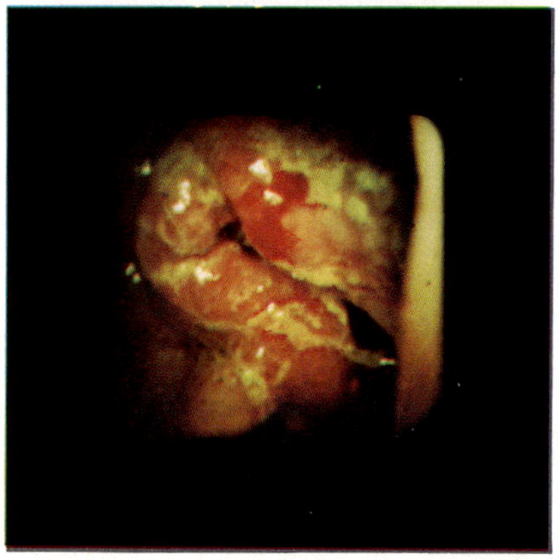

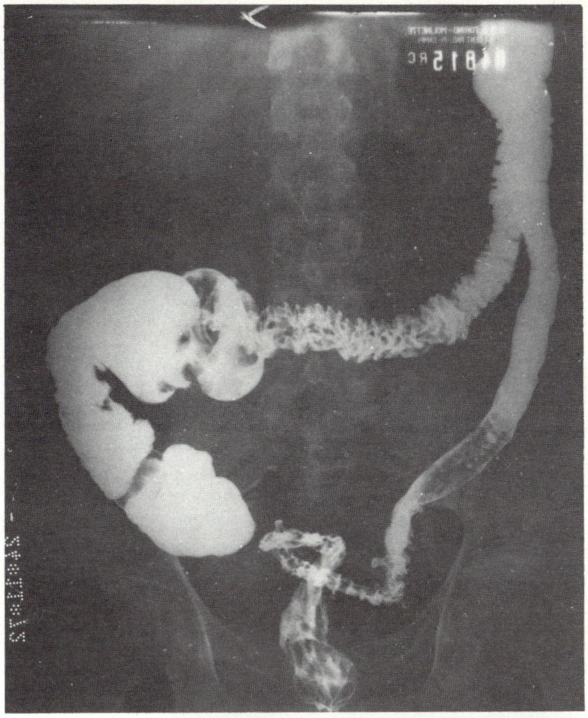

Fig. 375.
Barium enema. Annular narrowing of the transverse colon close to the hepatic flexure by an infiltrating tumour.

Fig. 376.
The fibrecoloscope arrested at the hepatic flexure.

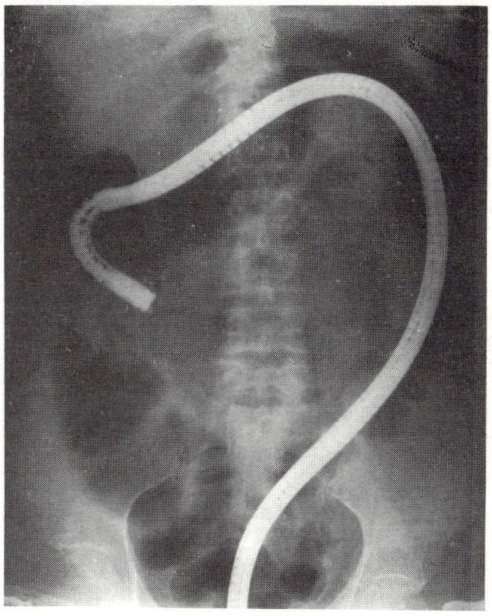

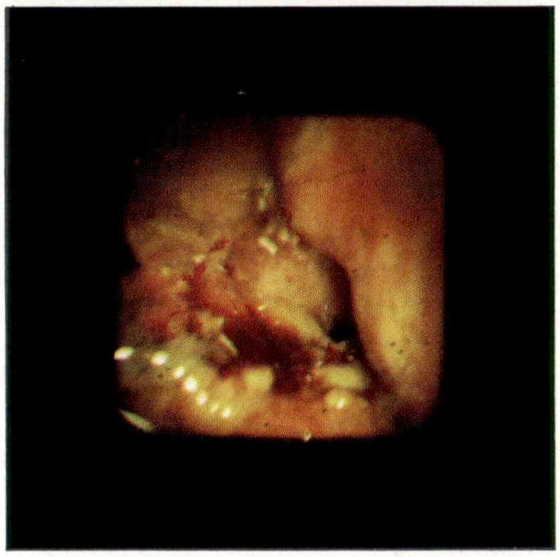

Fig. 377.
Infiltrating stenosing tumour. Ulcerated fungoid lesions infiltrating the bowel wall and stenosing the bowel lumen at the hepatic flexure. The diameter of the hepatic flexure is reduced to about 0.5 cm making further progression of the fibrescope impossible.

ig. 378.
lcerated stenosing fungoid tumour. bnormal endoscopic aspect of the ːcum which appears fixed and filtrated by extensive ulcerated, eeding fungoid lesions.

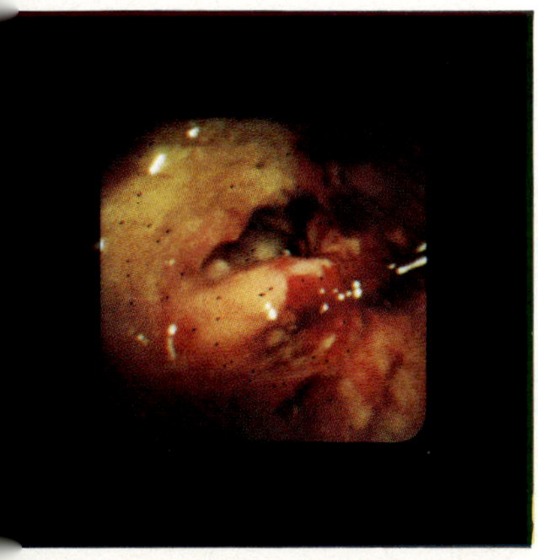

Fig. 379.
Surgical specimen. Ascending colon, cecum and last ileal loop. Gross ulcerated growths infiltrate the various layers of the bowel wall. Histologic diagnosis: adenocarcinoma.

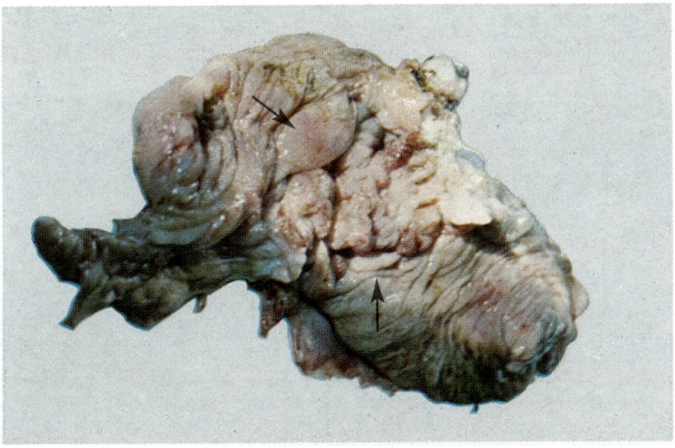

Fig. 380.
Tumoural stenosis. The mucosa of the splenic flexure is pinkish, smooth and shiny. The splenic flexure is fixed, rigid and infiltrated. Fungoid lesions can be seen through the patent bowel lumen.

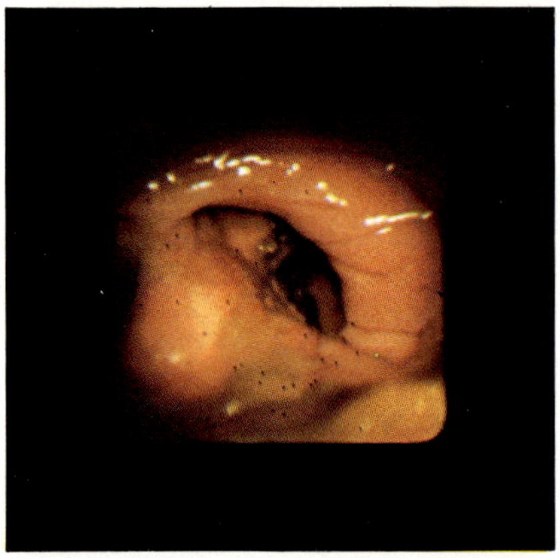

Fig. 381.
Stenosing fungoid tumour. Irregular fungoid lesion protruding in the distal portion of the sigmoid colon. The base of the tumour is wide and infiltrating and narrows the bowel lumen.

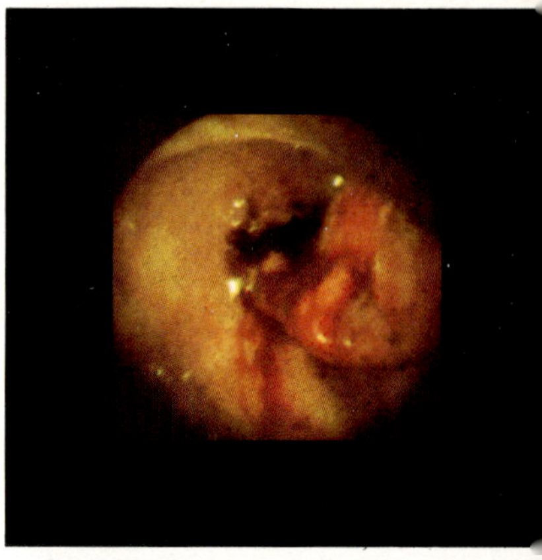

Fig. 382.
Stenosing fungoid tumour. Large, protruded, clapper-shaped growth in the descending colon. The base of the tumour infiltrates the bowel wall and narrows the bowel lumen.

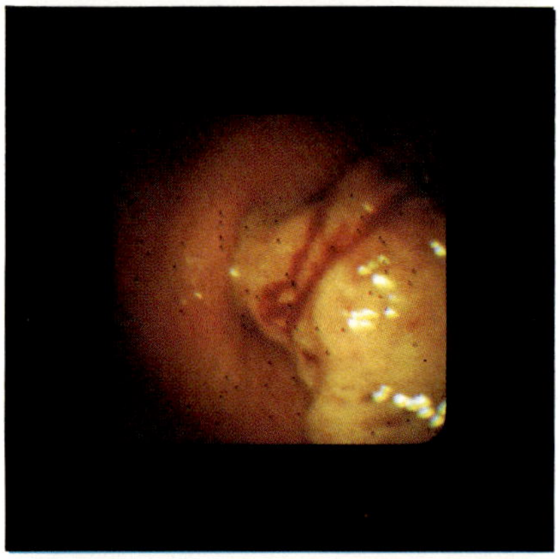

Fig. 383.
Stenosing tumour. Substenosis of the second portion of the sigmoid colon. The bowel wall is rigid and pitted with large ulcers with irregular borders. The bowel lumen (centre) is no wider than the head of a match.

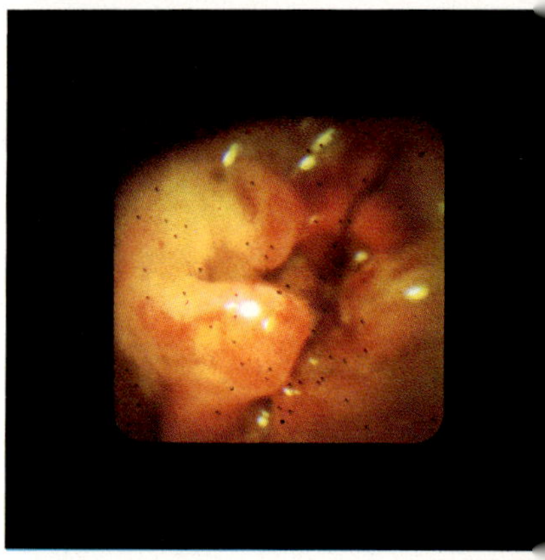

Fig. 384.
Stenosing fungoid tumour. Uneven-surfaced fungoid lesion covered with ulcerated, pinkish-red mucosa, stenosing the bowel lumen at the hepatic flexure.

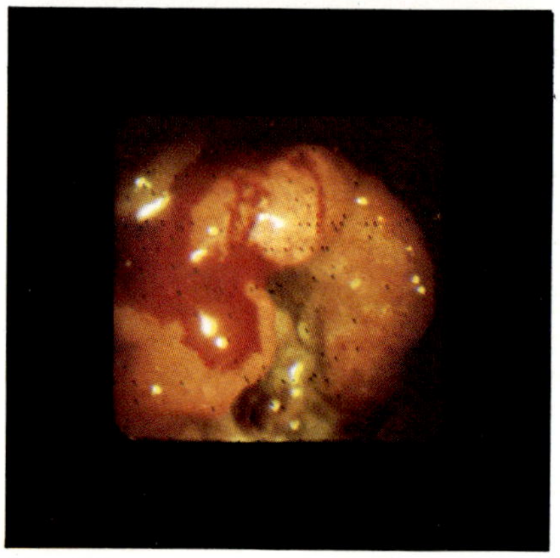

Fig. 385.
Stenosing infiltrating tumour. Massive infiltration of the bowel wall of the transverse colon by a smooth-surfaced tumour (see arrows) covered with pinkish-red mucosa indistinguishable from the unaffected mucosa nearby.

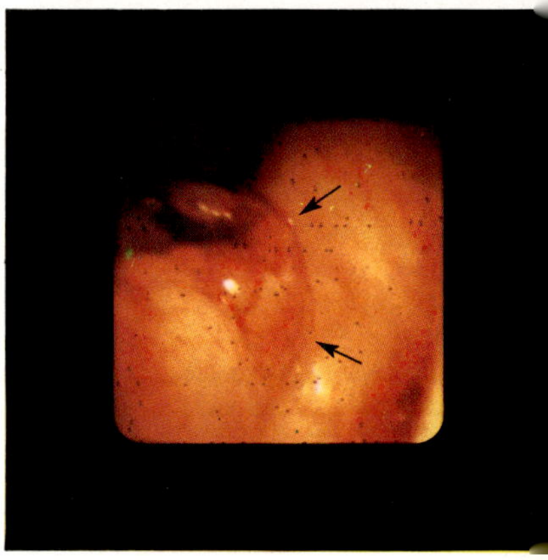

Fig. 386.
Stenosing ulcerated fungoid tumour. Piled-up growths of monstruous aspect infiltrating the bowel wall and stenosing the bowel lumen at the middle third of the transverse colon.

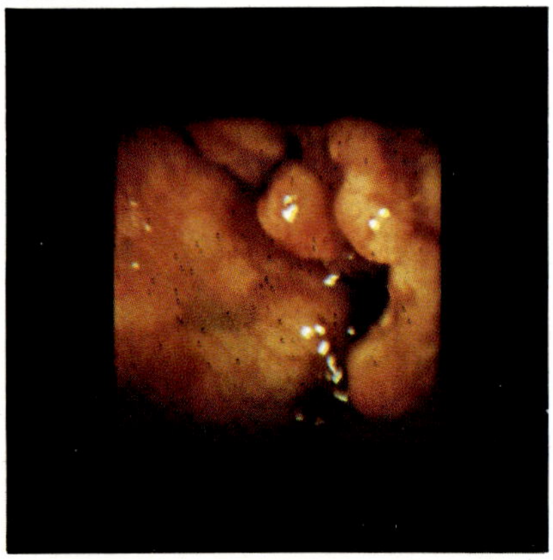

Fig. 387.
Close-up view of the same case. Through the narrowed lumen of the bowel it is possible to see that a portion of the wall of the colon several centimetres long is infiltrated and ulcerated.

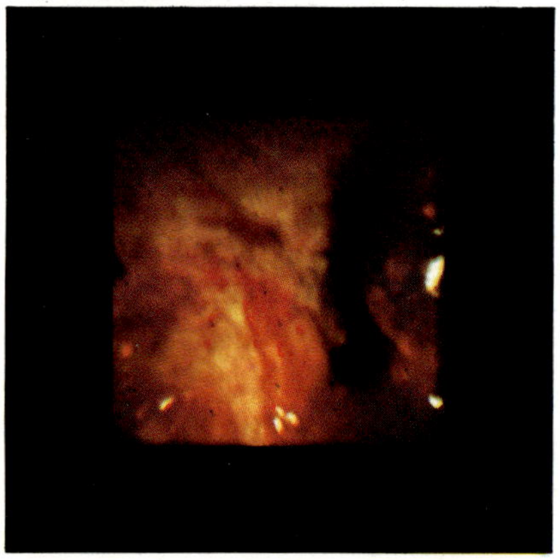

Fig. 388.
Surgical specimen. Part of the transverse and descending colon. Numerous cauliflower-like growths with central ulceration are visible on the mucosa. Histologic diagnosis: adenocarcinoma of the large bowel.

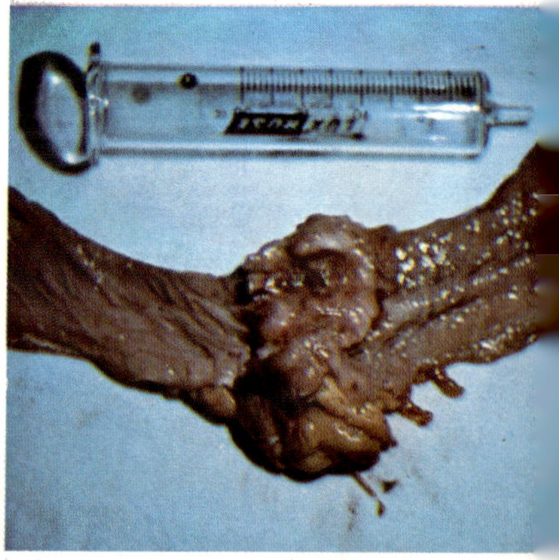

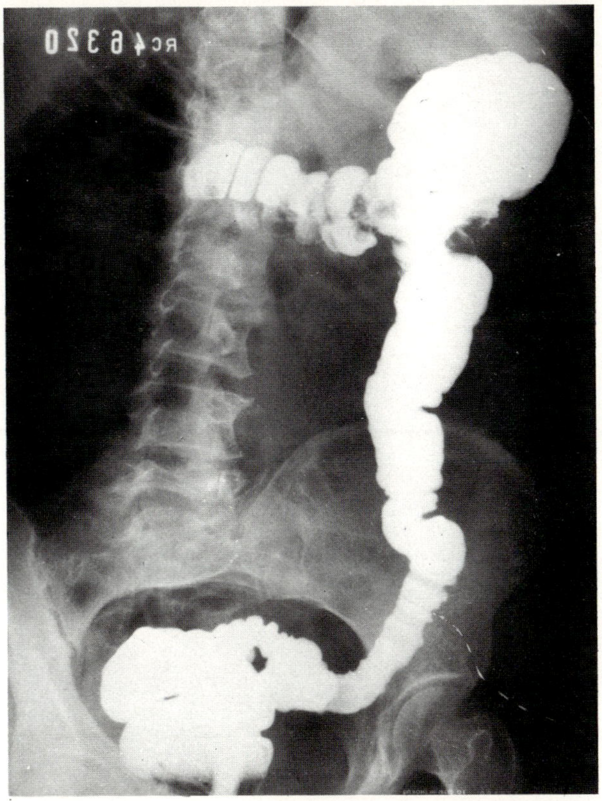

Fig. 389.
Barium enema. Irregular narrowing of the left colon close to the splenic flexure with a filling defect on its lateral aspect, corresponding to an infiltrating fungoid lesion.

Fig. 390.
Endoscopic view. Fungoid tumour. Large, broad-based fungoid growth with an irregular, ulcerated surface, covered with mucus and fibrin, at the left flexure of the colon. The nearby bowel wall is fixed and infiltrated.

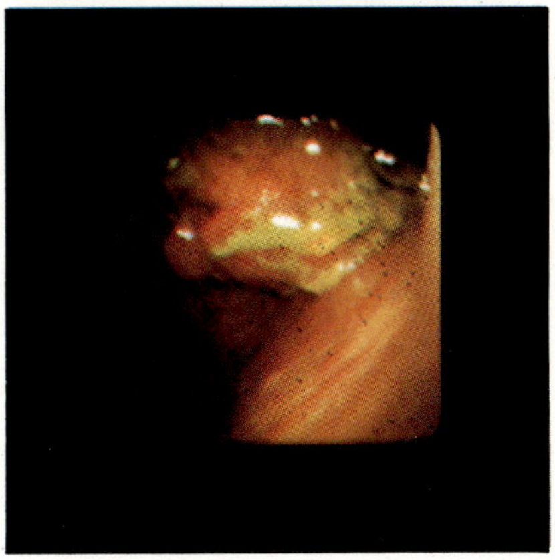

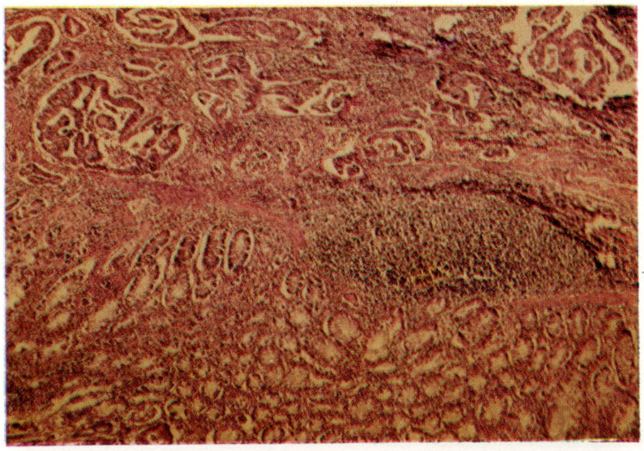

Fig. 391.
Coloscopic biopsy taken from the growth. Diagnosis: adenocarcinoma.

Fig. 392.
Surgical specimen. Part of the descending and transverse colon. A large neoplastic growth with an irregular, ulcerated surface of pinkish-red colour, the size of an orange, protrudes from the pinkish-red mucosa. Histologic diagnosis: adenocarcinoma.

Fig. 393.
Fungoid growths infiltrating the bowel wall at the sigmoid-descending colon junction.

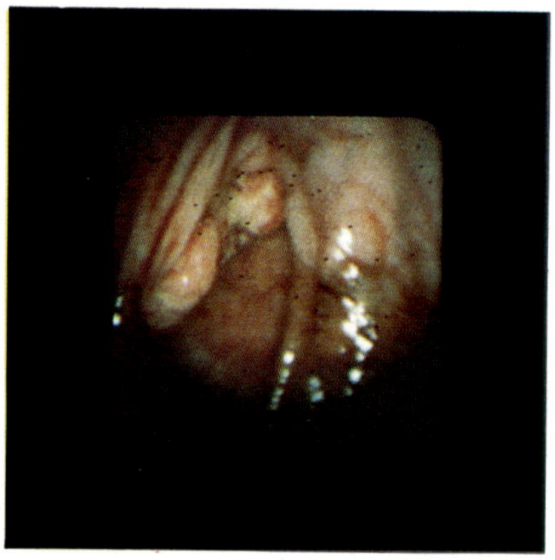

Fig. 394.
Closer view of the same case. Tough fungoid lesion with an irregular, ulcerated surface.

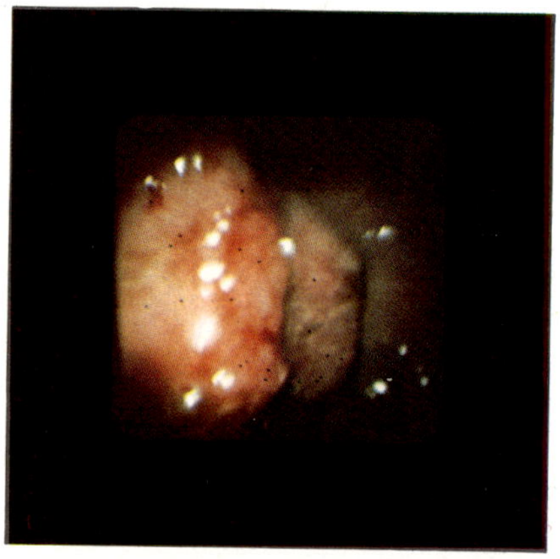

Fig. 395.
Coloscopic biopsy. Histologic diagnosis: adenocarcinoma.

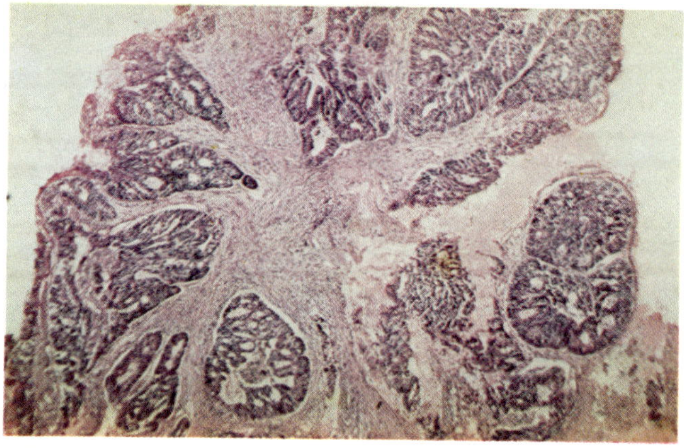

Fig. 396.
Surgical specimen. Sigmoid and part of the descending colon. Large cauliflower growth with an uneven surface of bright red colour. Histologic diagnosis: adenocarcinoma.

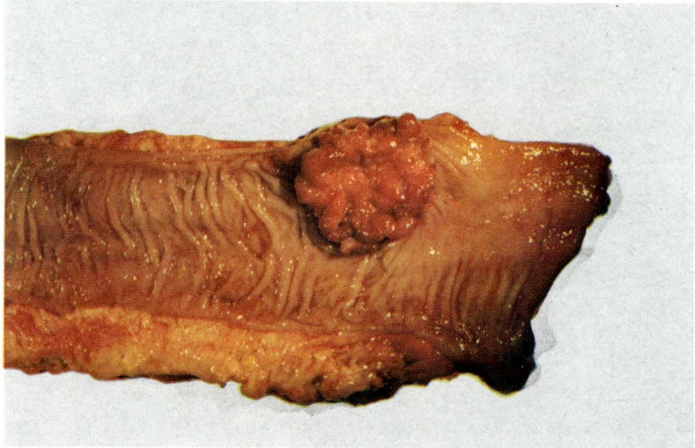

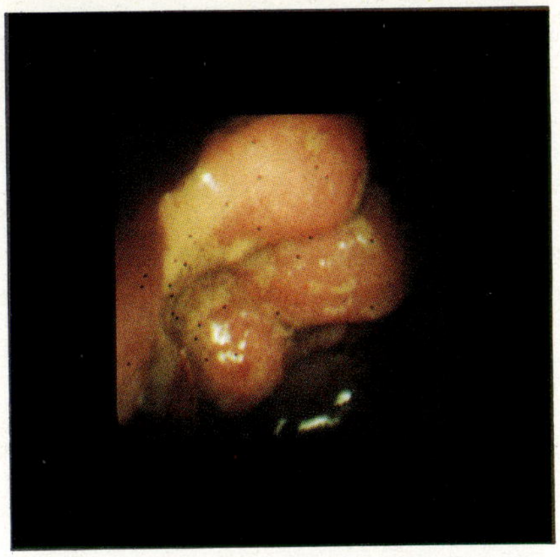

Fig. 397.
Polypoid cancer. Numerous polypoid growths protruding in the bowel lumen of the middle and distal portion of the sigmoid colon. The lesions have a smooth surface, with ulceration of the mucosa, and are partly covered with mucus and fibrin.

ig. 398.
olypoid cancer. Neoplastic growths hich seem to infiltrate the bowel all hidden under a fold of the escending colon.

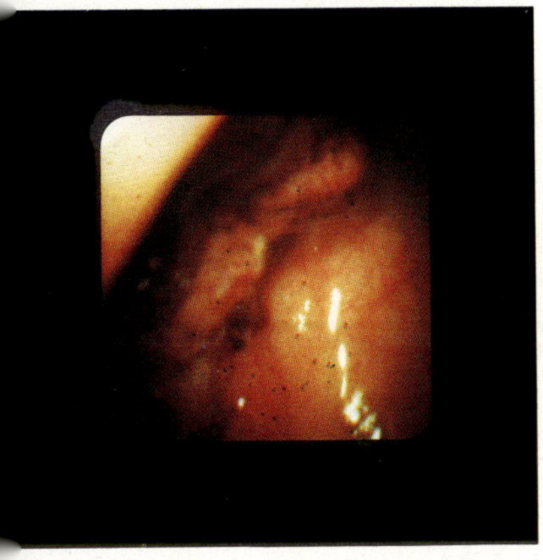

Fig. 399.
Coloscopic biopsy. Histologic diagnosis: adenocarcinoma.

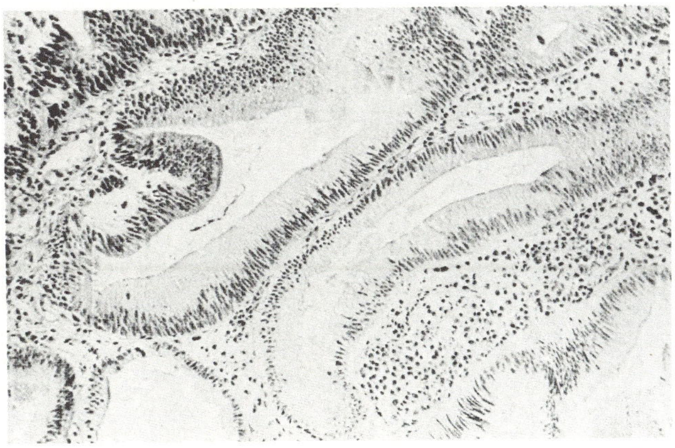

ig. 400.
urgical specimen. Sigmoid and escending colon. Long-stalked finely lcerated blackberry-like new rowths. Histologic examination of e stalk: mucosa of the large testine. Histologic examination of e tip: adenocarcinoma.

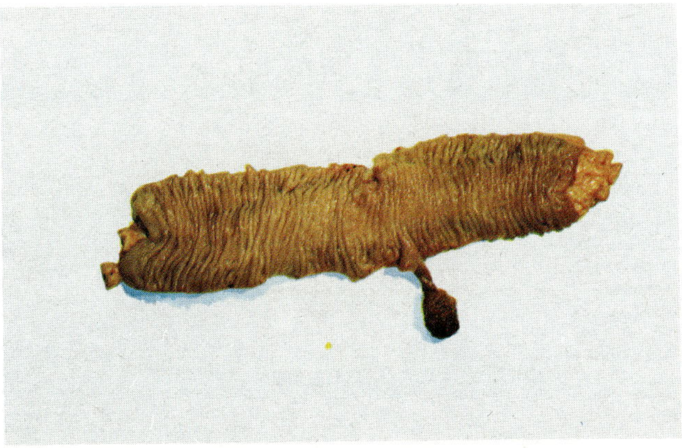

Extrinsic compression

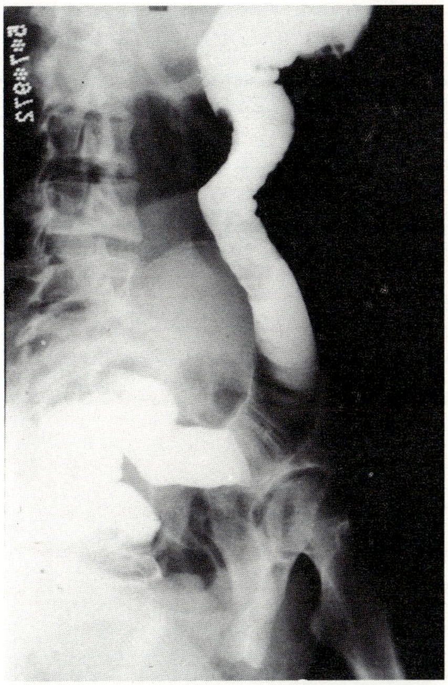

Fig. 401.
Barium enema. Circumscribed, medially concave, arch-shaped impression due to compression from without in the upper portion of the descending colon.

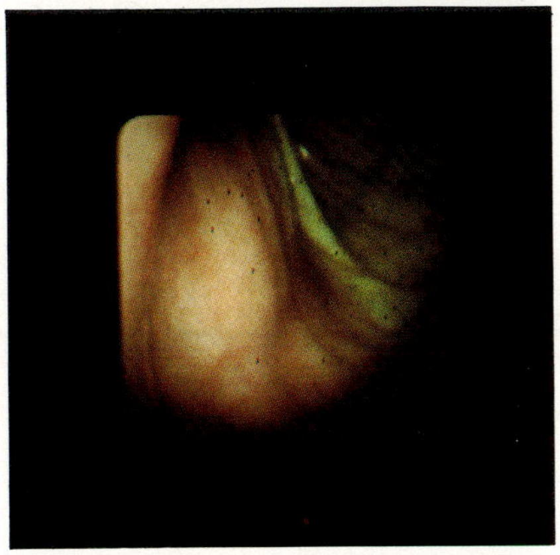

Fig. 402.
The radiologic Finding corresponds to a bulge covered with smooth, shiny, pinkish mucosa caused by compression from without. The bulge in the bowel wall, covered with elastic mucosa, can easily be displaced with the biopsy forceps.

Endoscopic aspects of surgical anastomoses

Endoscopic examination of surgical anastomoses is one of the major indications for coloscopy. Periodical inspection of the anastomosis after resection of the large bowel for carcinoma is essential for the early detection of possible recurrences at the anastomosis or in the cranial or caudal segments. Likewise, possible recurrences or complications which may be difficult to demonstrate radiologically in an early stage can be detected by coloscopic examination in patients subjected to ileotransversostomy for inflammatory conditions of the terminal ileum.

ɜnd-to-end anastomoses

g. 403.
ɹd-to-end anastomosis of the distal ɔrtion of the sigmoid colon. The ɪastomosis is unobstructed, elastic ɪd lined with greyish-pink, smooth, iny mucosa.

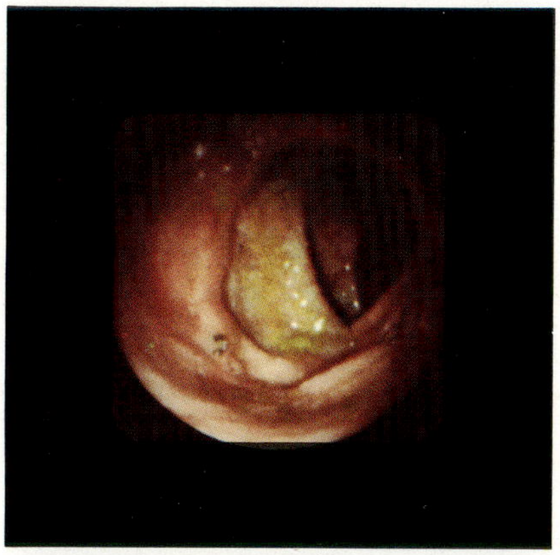

g. 404.
ɹd-to-end anastomosis. The ɪastomosis is unobstructed, elastic ɪd lined with pinkish, smooth ucosa. Along the anastomosis there e numerous elastic elevations vered with pinkish-red mucosa tributable to granulation tissue.

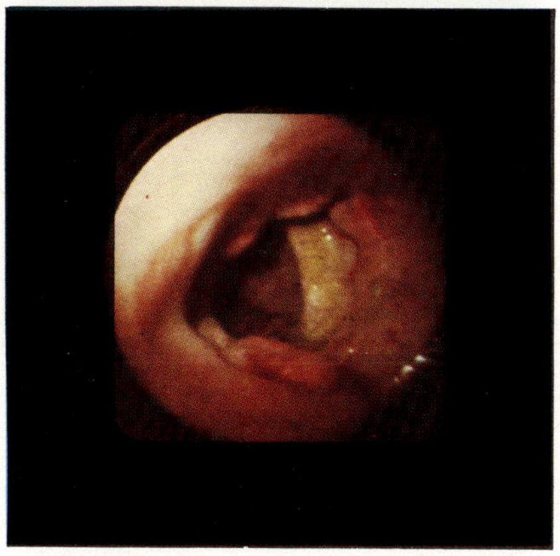

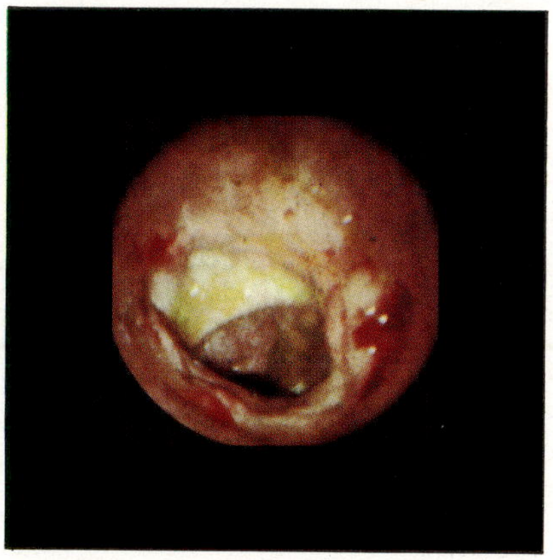

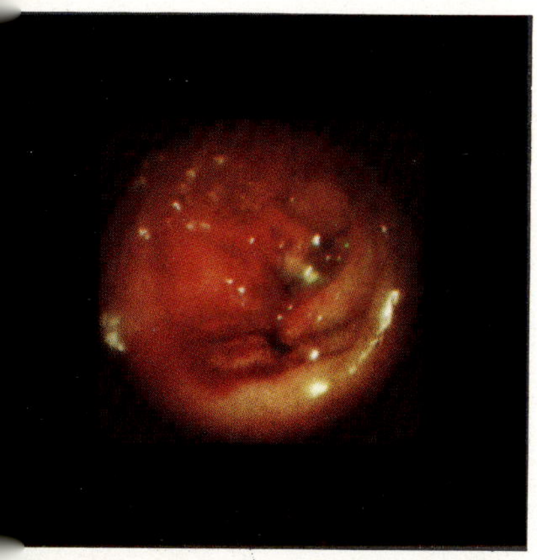

Fig. 406.
Recurrence of carcinoma on an end-to-end anastomosis. The opening of the anastomosis is not very distensible on insufflation, fixed and stiff. The lumen of the anastomosis is narrowed by large infiltrated folds covered with sanious mucosa. Histologic examination revealed an adenocarcinoma.

Fig. 405.
Same case. Biopsy on the peri-anastomotic excrescences. Bleeding after biopsy is slight. Histologic examination confirmed the endoscopic conclusions (regenerating mucosa).

suture remnants

Fig. 407.
End-to-end anastomosis with suture remnant. Attempts to remove the stitch which is surrounded by an excess of granulation tissue causes fairly severe bleeding.

Fig. 408.
The surgical anastomosis can only be distinguished on account of the presence of suture remnants. There is intense oedema and congestion of the mucosa around the anastomosis. The lumen is partly obstructed by large folds covered with smooth, shiny, wine-red-coloured mucosa. Numerous stitches incarcerated between the inflamed folds can be distinguished. There is no evidence of a neoplastic recurrence.

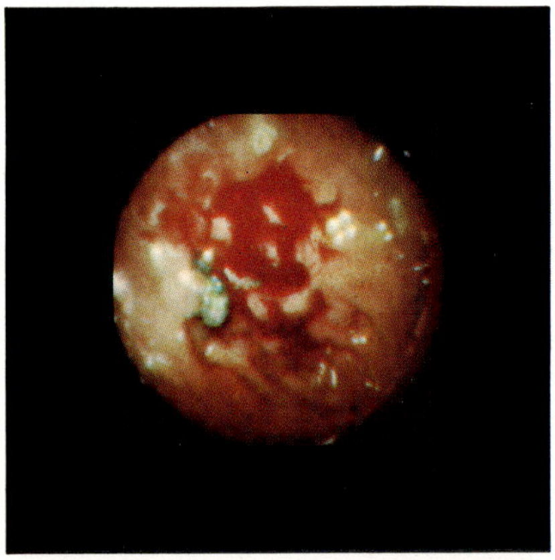

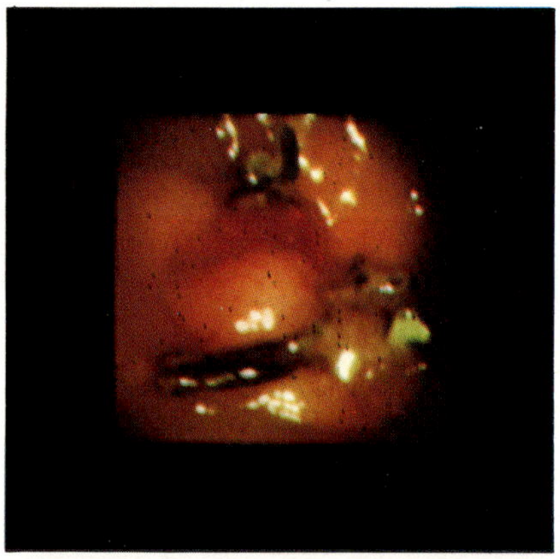

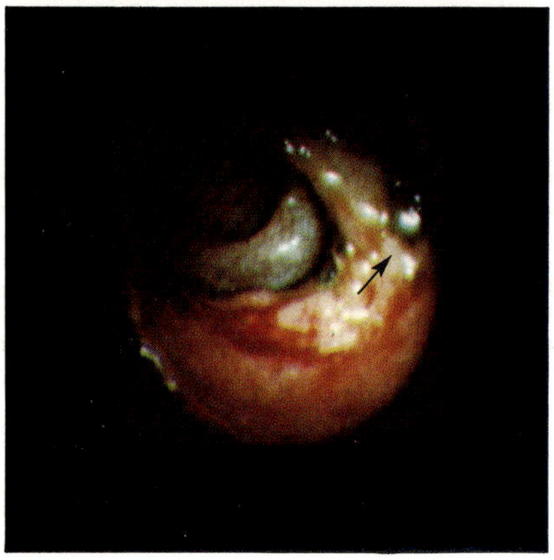

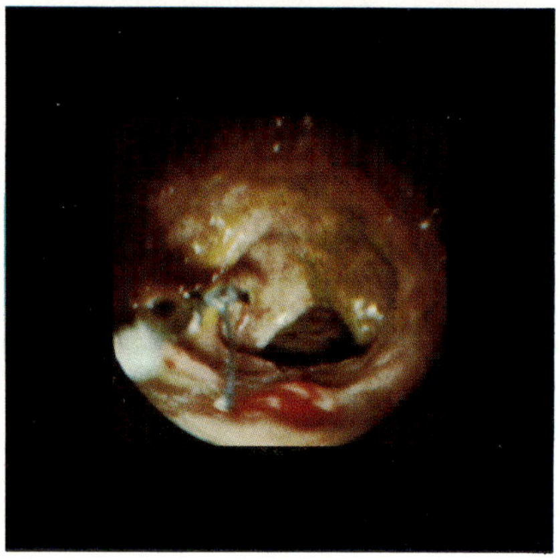

Fig. 409.
End-to-end anastomosis with suture remnant.

Fig. 410.
Same case. The suture is easily removed with the biopsy forceps.

leotransversostomy

g. 411.
otransversostomy. The
astomosis is distensible and
nctionally efficient. The mucosa is
nkish and smooth. A tangle of
otruding surgical stitches can be
n, part of which were removed
th the biopsy forceps.

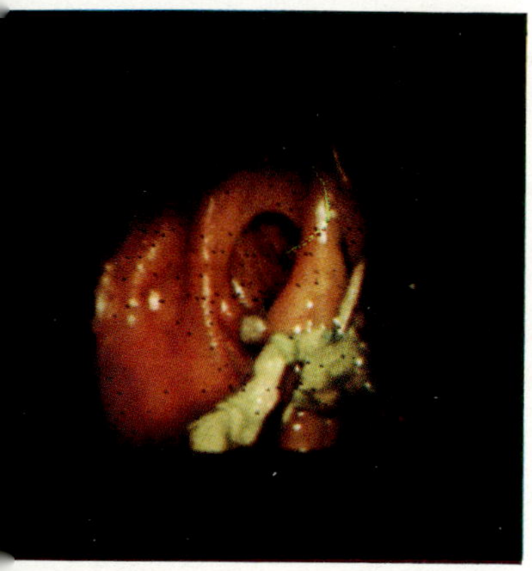

Fig. 412.
Endoscopic view of an ileotransversostomy. The surgical anastomosis is open. The mucosa is congested, oedematous, brittle and bleeds easily on contact with the endoscope.

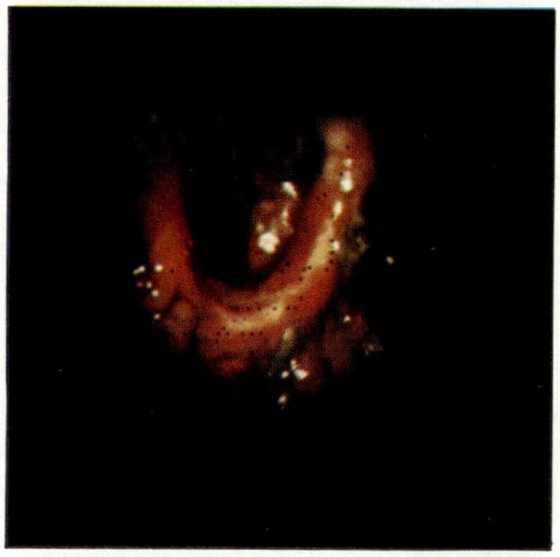

Fig. 413.
Detail of the same case. Numerous ulcers can be seen on the colic side of the anastomotic opening. A suture remnant can also be seen in this case.

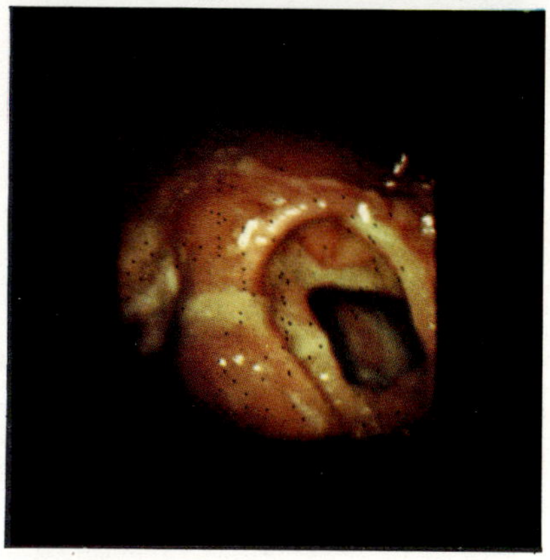

Fig. 415.
Endoscopic view of the ileotransversostomy. The anastomosis is rigid, infiltrated and lined with intensely congested mucosa with patches of mucus and fibrin. Flat-bordered ulcers containing a fibrinous material are scattered over the ileal mucosa for a length of about 5 cm. Histologic examination of specimens obtained with the biopsy forceps from various places revealed a recurrence of Crohn's disease of the anastomosed ileal loop.

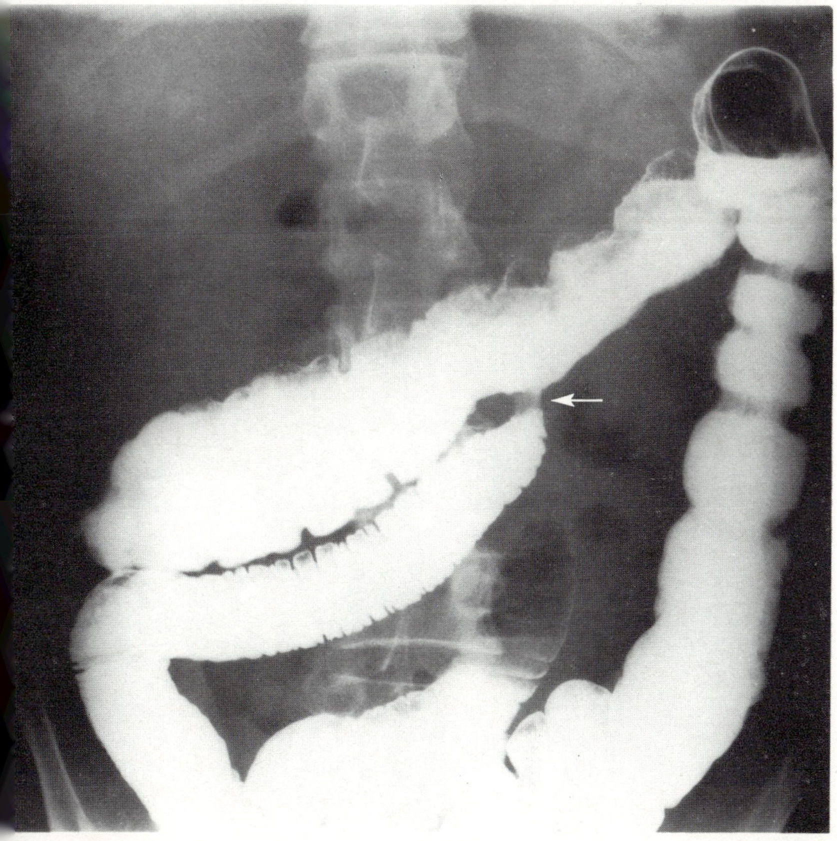

Fig. 414.
Barium enema. Right colectomy with ileotransversostomy. The anastomosis is open but narrowed, rigid and apparently infiltrated. The adjacent colon has an irregular, wavy contour.

Part three

Sigmoidoscopy with the fibrescope in a case of new-formed rectal bladder

'he fibrecoloscope can be used to inspect the rectum and
.gmoid colon isolated to replace the urinary bladder after total
ystectomy.
n this case the ureters were implanted into the sigmoid colon.
/ith the aid of the coloscope it was possible to examine the
olated rectosigmoid cavity and to investigate the function of
ne ureters.

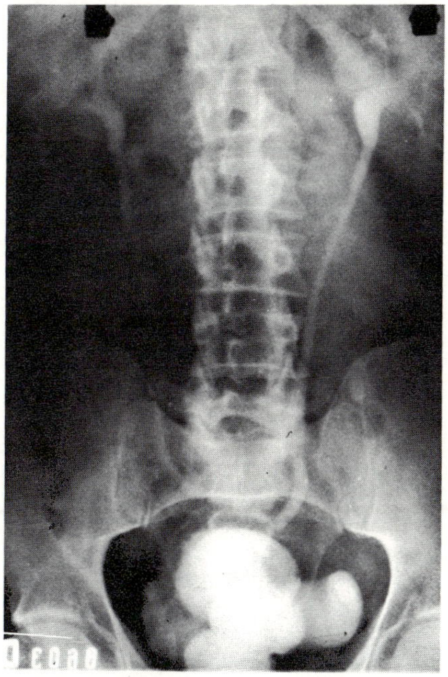

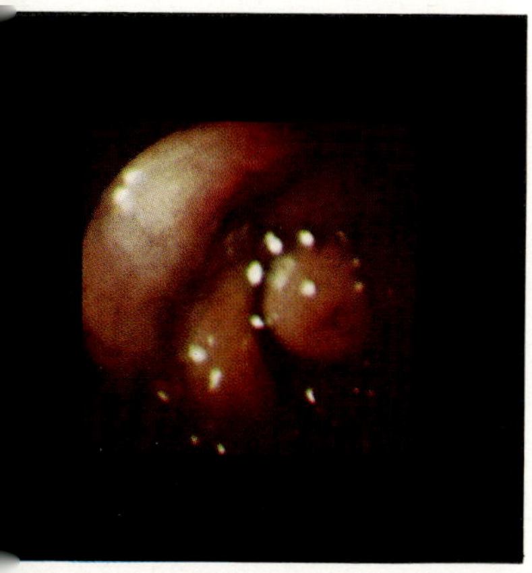

Fig. 416.
At urography the size and position of the ureters are normal. They are implanted in the middle part of the sigmoid colon. There is no obstruction to the flow of the radio-opaque urine into the new-formed bladder.

Figs. 417 & 418.
Endoscopic view of the outlet of the ureters after an intravenous injection of methylene blue.

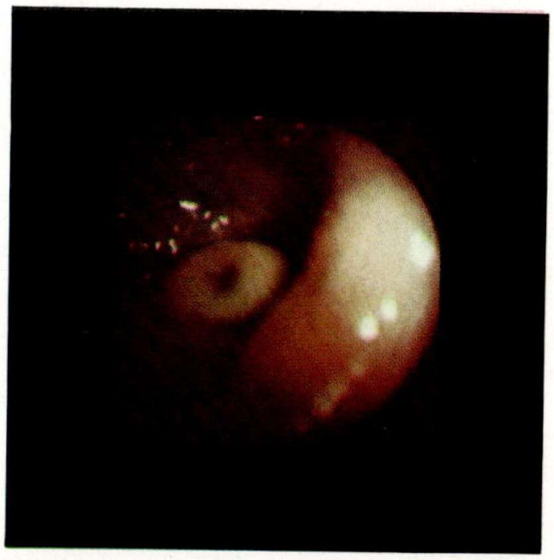

Emergency coloscopy

mergency coloscopy has proved a valuable and decisive iagnostic procedure in 60-70% of cases of rectal bleeding. ny patient admitted to hospital with severe low digestive act bleeding should immediately be subjected to endoscopy nless shock is present in which case the examination should e put off for a day or two. Emergency coloscopy requires no owel preparation because the bleeding will have effectively leared the colon of faeces. The presence of faeces unmixed ith blood indicates that the coloscope has passed beyond the te of the bleeding. No premedication is required. The ethod of examination is the same as for normal coloscopy. he main difficulties which may be encountered are due to the resence of blood on the bowel wall, which obstructs the view f the mucosa, and, especially, to the presence of clots, which annot always be removed. The mucosa can be visualized by epeatedly washing out with cold water injected through a eflon tube inserted through the biopsy forceps channel.

ndications: all cases of severe low digestive tract bleeding.

ontraindications: severe shock.

recautions:

- the bowel lumen must always be in view;
- air must be insufflated sparingly, just enough to distend the bowel wall, to avoid the risk of increasing bleeding by excessive distension;
- traumatic manoeuvres must be avoided;
- the pulse rate and blood pressure must be checked frequently;
- fluoroscopy must be used;
- the examination must be discontinued if bleeding ulcerated diverticula are found.

mergency coloscopy has the following advantages:

- it can be performed in the presence of bleeding;
- it can show the site and, often, the cause of bleeding in a high proportion of cases;
- it will not aggravate the patient's conditions and can therefore be performed with little danger;
- it will not increase bleeding in the hands of an experienced examiner provided adeguate precautions are taken and continuous visual control is maintained;
- it may rule out the need for explorative laparatomy;
- it can indicate the correct medical or surgical treatment by demonstrating the site and cause of bleeding.

oloscopy performed two days after an episode of acute rectal eeding ("delayed emergency").

this case coloscopy was decisive, showing the cause of the eeding and making it possible to perform aimed biopsy and us to make an early diagnosis of carcinoma.

Figs. 419, 420, 421 & 422.
Emergency coloscopy. Bleeding ulcerated cancer of the sigmoid colon. Large ulcerated, bleeding cauliflower growths are clearly visible in these four figures. Fresh blood covers the growth in Fig. 419 making it impossible to discern the origin of the bleeding.

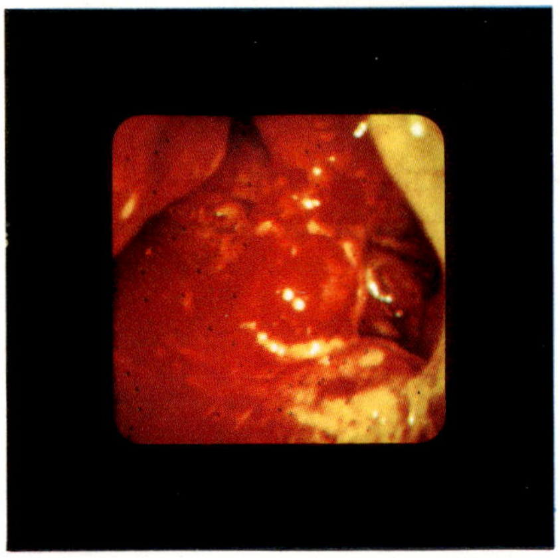

Fig. 419.

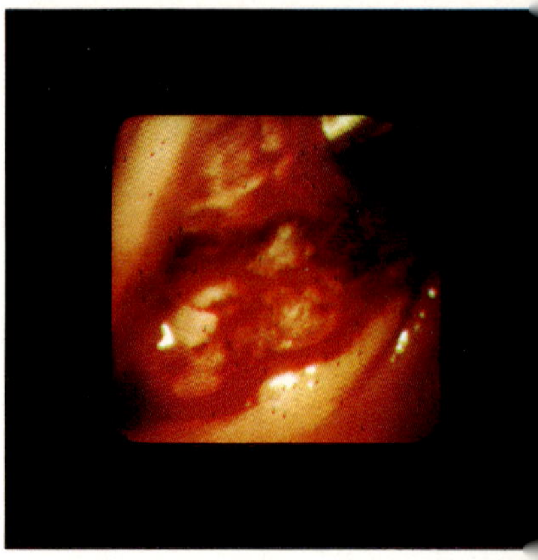

Fig. 420.

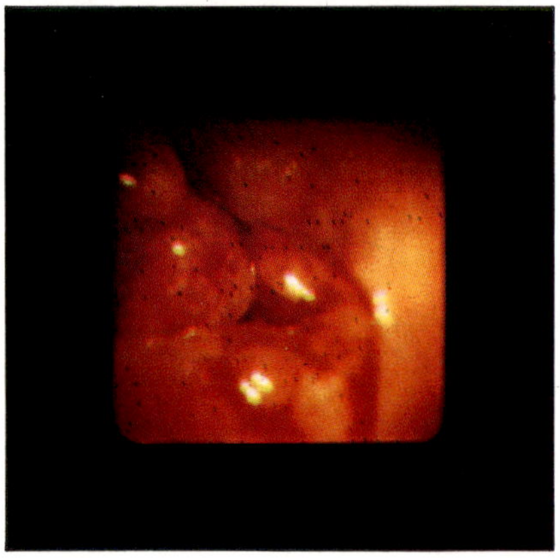

Fig. 421.

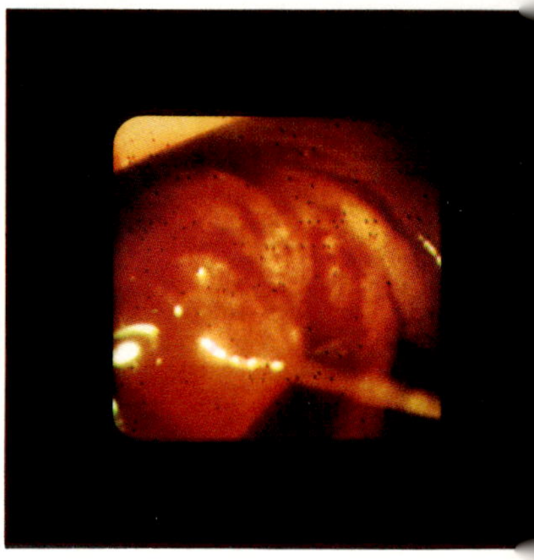

Fig. 422.

Fig. 423.
Emergency coloscopy. Intestinal bleeding in a case of active chronic idiopathic ulcerative rectocolitis. The wall of the large bowel is coated with bright red blood due to oozing haemorrhages. Passage of the endoscope aggravates the bleeding. Widespread anatomical changes of the bowel are apparent: oedema of the mucosa, bleeding punched-out ulcerations, pseudopolyps.

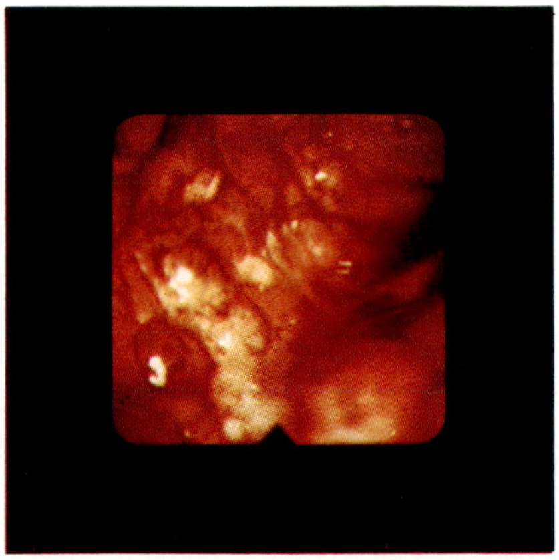

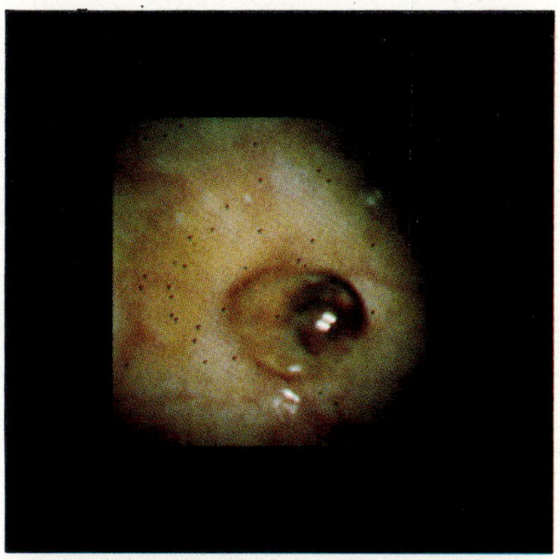

Fig. 424.
Small sessile new growth of the distal portion of the descending colon. Clotted blood on the tip of the growth.

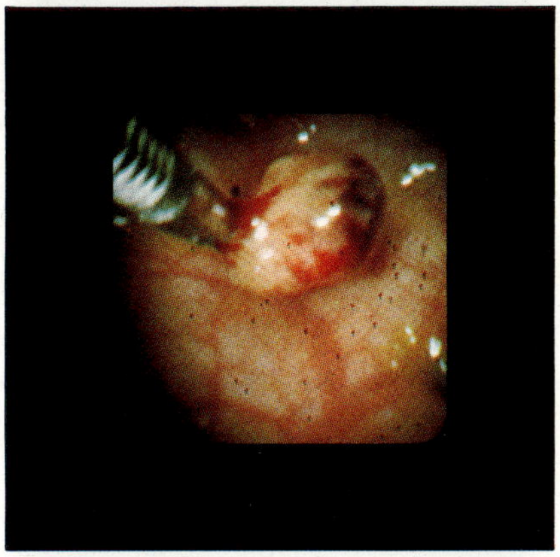

Fig. 425.
Biopsy at the base of the new growth.

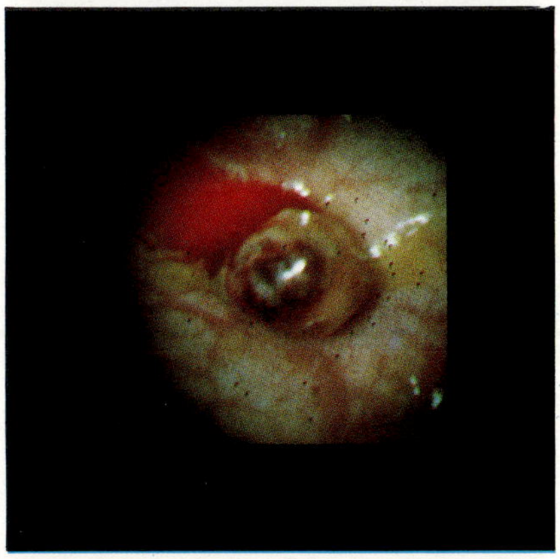

Fig. 426.
Bleeding after biopsy.

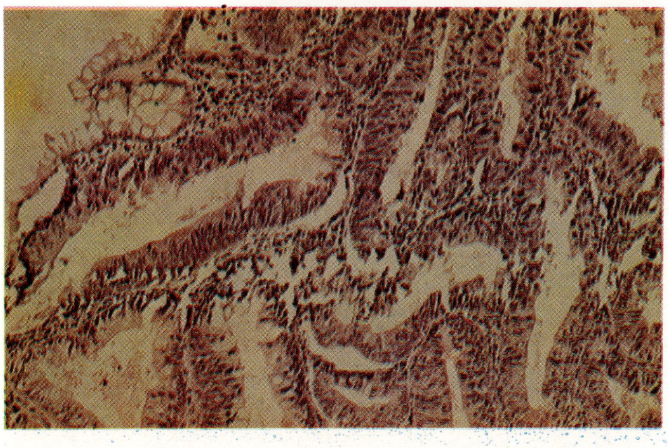

Fig. 427.
Coloscopic biopsy. Histologic preparation: adenocarcinoma of the large bowel.

Figs. 428, 429, 430, 431, 432 & 433. Emergency coloscopy.
Endoscopic view of various sections of the large bowel from the second portion of the sigmoid colon to the ascending colon, ileocecal papilla (Fig. 432) and last loop of the ileum (Fig. 433).

The wall of the large bowel is uniformly daubed with fresh blood. No faeces are present. The mucosa is smooth and no ulcerative lesions are visualized by repeated lavage of the mucosa at different levels. Bright red blood is also seen descending from above in the last ileal loop.
In this case emergency coloscopy and gastroduodenoscopy made it possible to establish that the site of bleeding was in the sub-duodenal portion of the small intestine. At surgery small multiple angiomas were found in the small intestine.

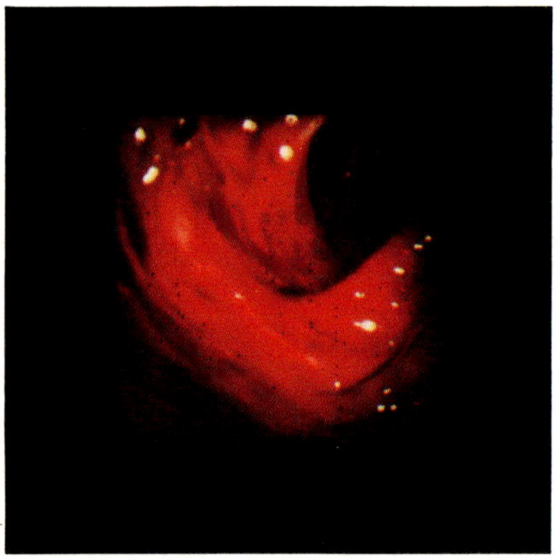

Fig. 428.

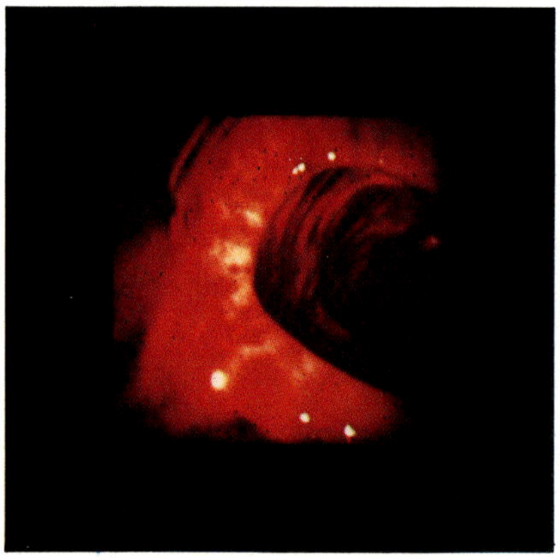

Fig. 429.

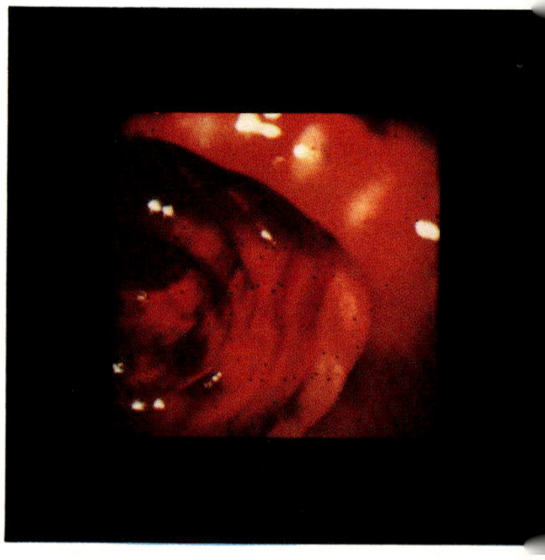

Fig. 430.

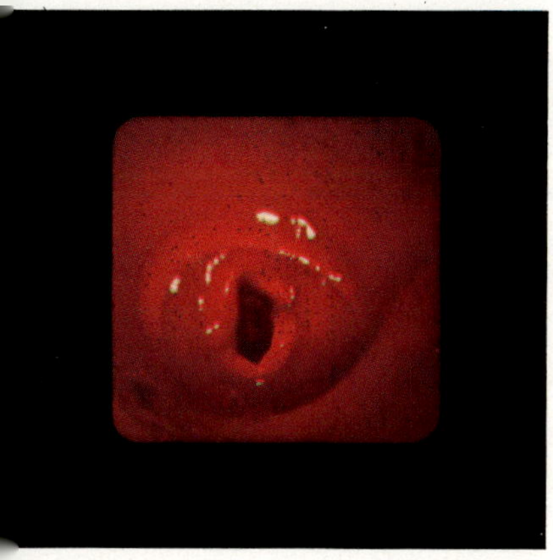

ig. 431.

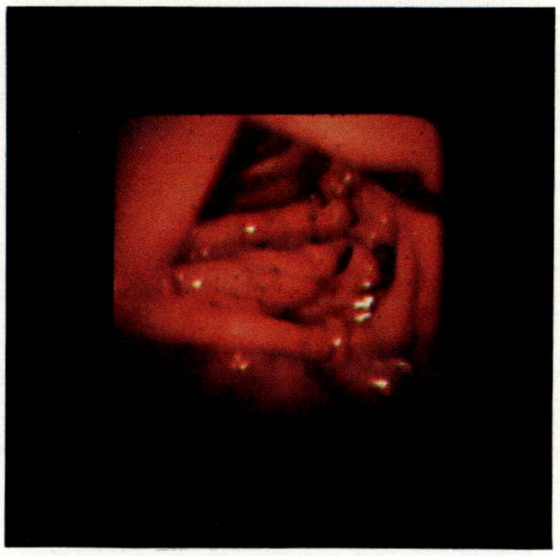

Fig. 432.

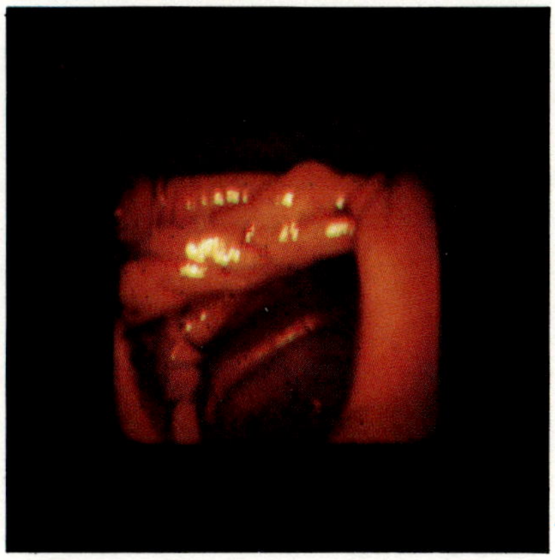

Fig. 433.

Fig. 434.
Emergency coloscopy. Ulcerated bleeding diverticula. The mucosa of the second portion of the sigmoid colon is very congested and oedematous. A group of ulcerated bleeding diverticula can be seen. The diverticula were covered with blood and were visualized by washing out with cold water.

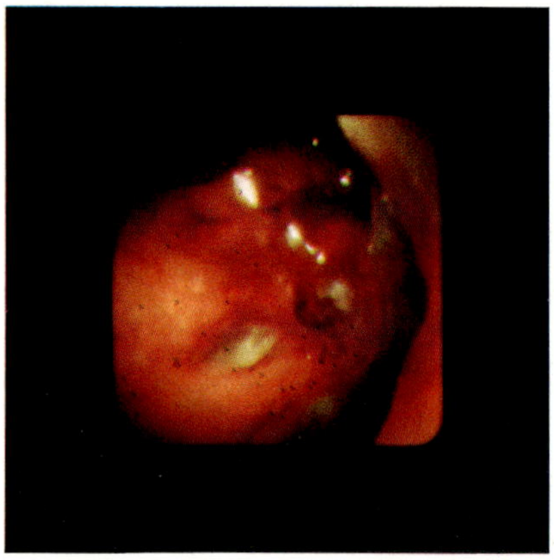

Fig. 435.
Emergency coloscopy. Numerous diverticula are visible in the descending portion of the sigmoid colon. Fresh blood is present in the lumen of the sigmoid colon. The figure shows a diverticulum with a round opening which is clearly bleeding. (Courtesy of dr. Armengol-Miro, Barcelona.)

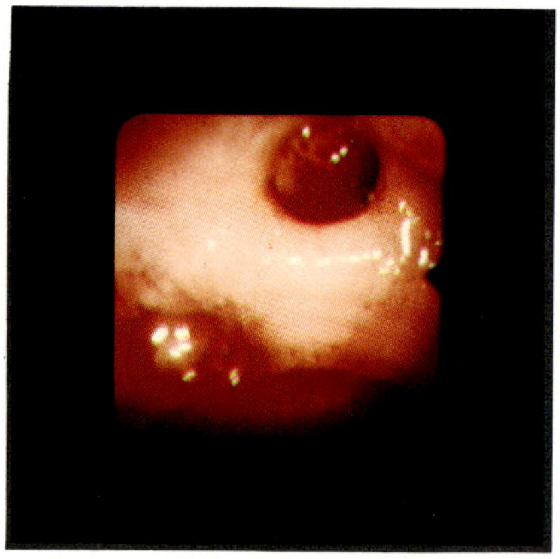

ïig. 436.
:mergency coloscopy. Transient schaemic colitis. The walls of the arge bowel at the sigmoid-lescending colon junction are coated vith fresh blood. Linear ulcers with leeding borders are scattered over bout 15 cm of the large bowel; bove this area the mucosa is ormal.

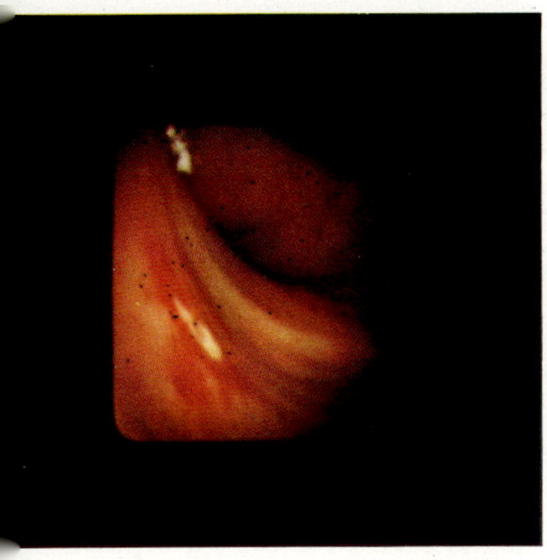

Fig. 437.
Same case. Numerous ulcers with regular, flat, bleeding borders are grouped together higher up in the descending colon.

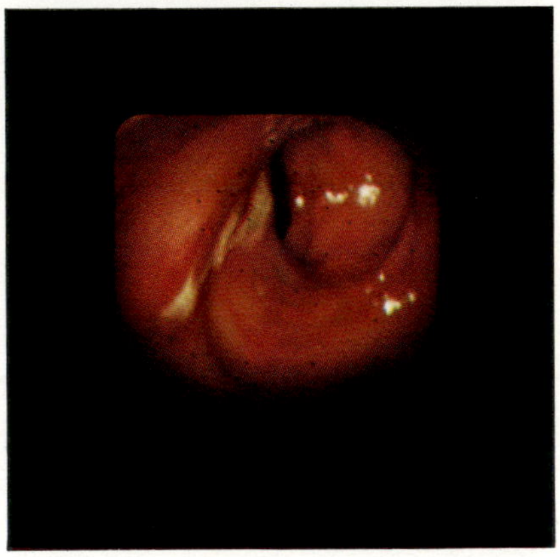

Fig. 438
Control coloscopy after six days of medical treatment. The mucosa of the colon is pinkish-red, shiny, and forms elastic folds of regular thickness. The previously observed ulcers are no longer visible and have been replaced by areas of intensely hyperaemic mucosa.

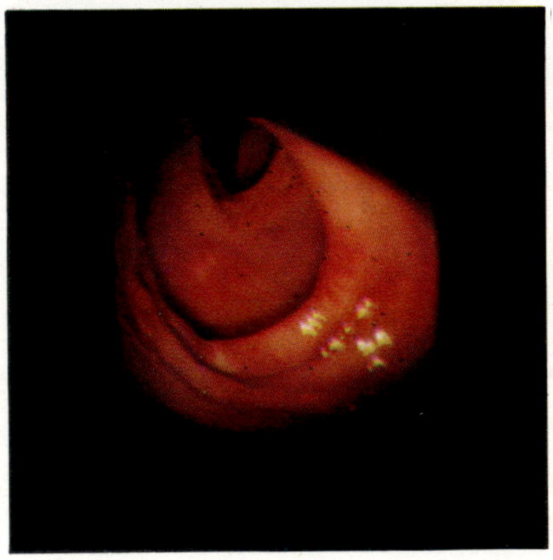

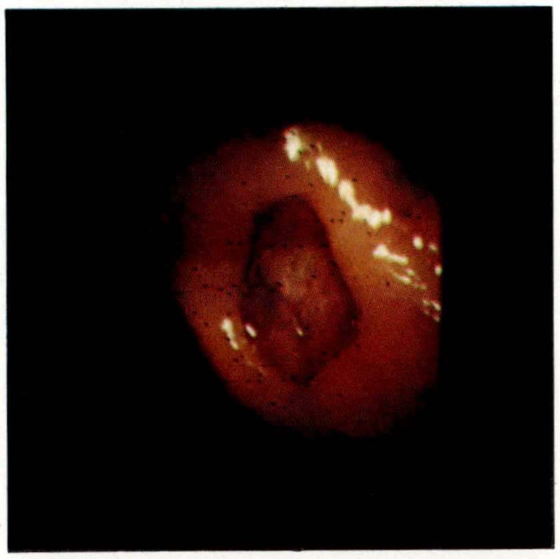

Fig. 439
Emergency coloscopy. Ischaemic colitis. Deep ulcer with haemorrhagic, necrotic material in the distal portion of the sigmoid colon.

Peroperative coloscopy

by A. Montori*

* Digestive Endoscopy Centre, 3rd Surgical Clinic
(Chief: prof. G. Marcozzi), University of Rome Medical School.

Peroperative coloscopy is coloscopy done in conjunction with laparotomy. Progression of the coloscope is aided by manipulation of the bowel by the surgeon who can thus steer its tip to a doubtful area under simultaneous visual control by the endoscopist. Important diagnostic problems can easily be solved with this technique which now represents a further application of gastrointestinal endoscopy. Although present flexible fibreoptic endoscopes will usually easily negotiate the various obstacles and bends strewn along the gastrointestinal tract, some parts of the large bowel can sometimes only be reached with great difficulty and lesions in these areas are not easily detected. There may be difficulty in passing the flexures and other parts of the large bowel to reach the terminal ileum. The two main advantages of peroperative coloscopy are:

– the surgeon can examine the bowel with transillumination which shows up the fine vascular networks and possible shadows due to endoluminal or intramural lesions;
– the endoscopist can easily reach any part of the large bowel and thus supply the surgeon with exact information on the pathological anatomical conditions there.

Peroperative coloscopy is therefore specifically indicated for investigation of those parts of the large bowel which cannot easily be visualized preoperatively and especially in cases in which the preoperative endoscopic findings are equivocal (bleeding of uncertain origin or minimal lesions which may be hard to detect during preoperative coloscopy on account of the bends along the colon). It can be used, even during emergency operations, to determine the exact site and extension of lesions of the large bowel and thus avoid a viscerotomy, and also makes it possible to establish the extension of surgical demolition in advance. Peroperative coloscopy entails no specific hazard provided it is done by a skilled endoscopist in close collaboration with the surgeon.

Fig. 440.
Peroperative endoscopy.
Position of the surgeon and endoscopist.

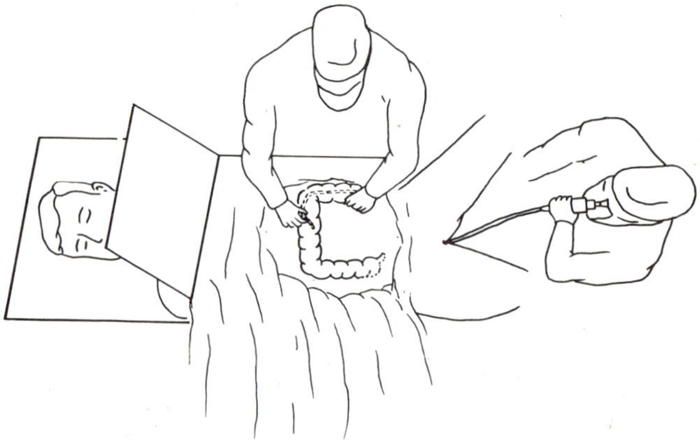

Fig. 441.
Peroperative endoscopy.
Radiological screening. The fibrecoloscope has reached the right colon.

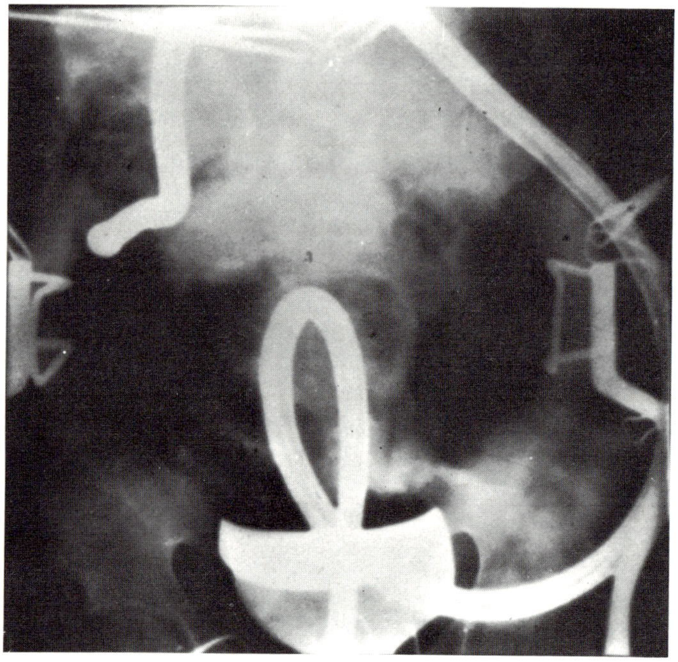

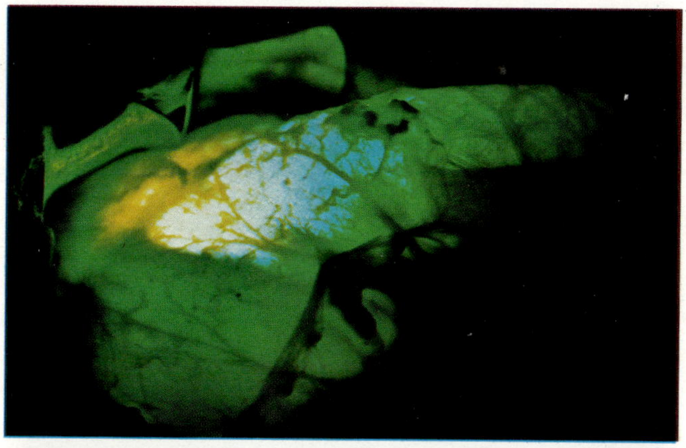

Fig. 442.
Peroperative coloscopy.
Infrared ray photograph.
Transillumination of the transverse colon showing the fine vascular networks of the bowel wall.

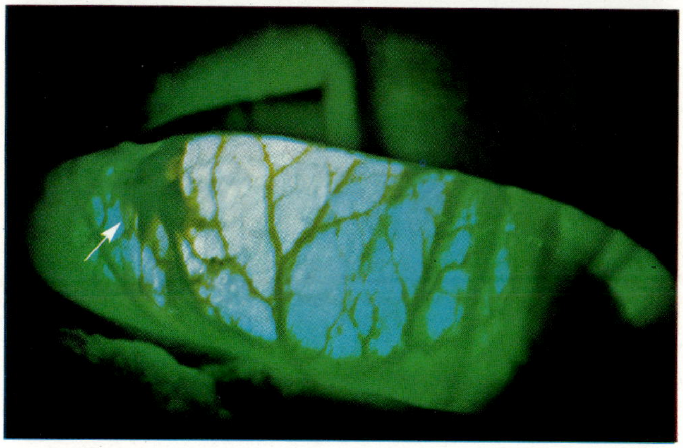

Fig. 443.
Peroperative coloscopy.
Transillumination of the descending colon. The arrow points to polyps in the bowel lumen.

Figs. 444, 445, 446 & 447.
Peroperative coloscopy.
View of the bowel wall of the descending and transverse colon with transillumination.
The fine vascular networks of the colon and the shadows of many small and large polyps are clearly visible.
Small lesions of the colon cannot always be detected by palpation alone. Simultaneous endoluminal inspection by the endoscopist and external inspection and palpation by the surgeon make it possible to detect, locate and remove all the growths.

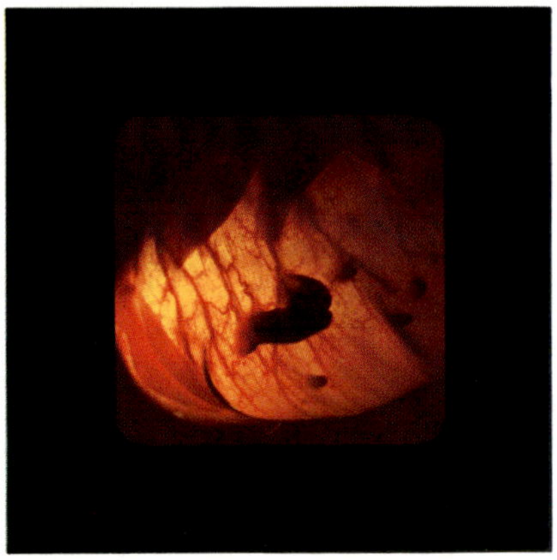

Fig. 444.

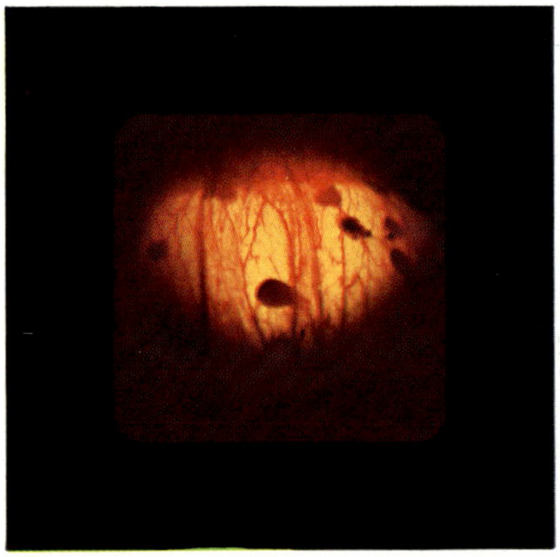

Fig. 445.

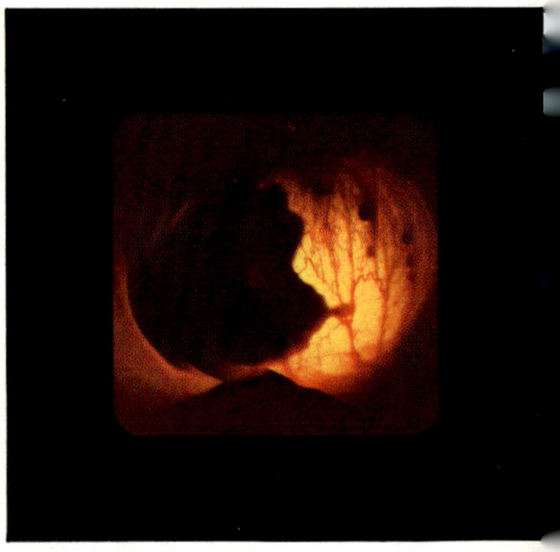

Fig. 446.

Fig. 448.
Peroperative coloscopy.
Endoluminal view of the descending colon after laparotomy. Sessile and pedunculated polyps of various sizes and the fine vascular network of the bowel wall distended by the insufflated air can clearly be distinguished.

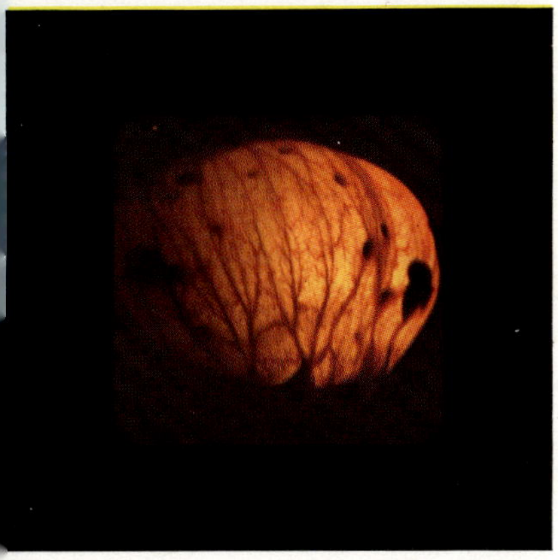

Fig. 447.

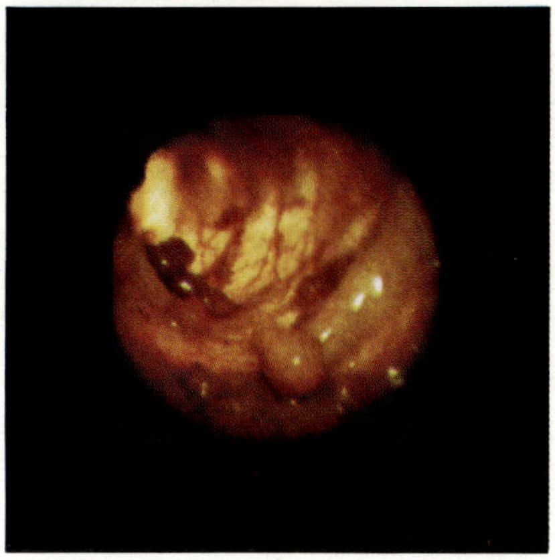

Coloscopic polypectomy

With the addition of suitable accessories the coloscope can be used to remove pedunculated or sessile polyps of the colon, as well as for diagnostic purposes.

advantages of coloscopic polypectomy

1) removal and retrieval of the polyp;
2) correct classification of the excised growth by means of serial histologic preparations;
3) prevention of the complications of polyposis (bleeding, myxorrhoea, colic spasm, tenesmus, irregular bowel function, cancer of the large intestine);
4) avoids surgery (laparotomy and viscerotomy);
5) can be done on an outpatient basis or with 48 hours hospitalisation;
6) requires no anaesthesia, being painless. (In anxious, hypersensitive subjects a sedative can be given - Valium, 10 mg by vein or Dolantin, 1 ampoule by intramuscular injection.);
7) has few contraindications (diverticulosis, large sessile growths infiltrating the wall of the colon, haemorrhagic diseases);
8) financial, social and psychological advantages.

equipment

Long or short fibrescope.
Diathermy snares of various types.
High frequency electricity source.
Nitrogen or carbon dioxide gas.
Indifferent plate. By-pass connections.
X-ray table for fluoroscopy.

Fibrescope
Any type of fibrecoloscope can be used to remove endoluminal growths of the colon.

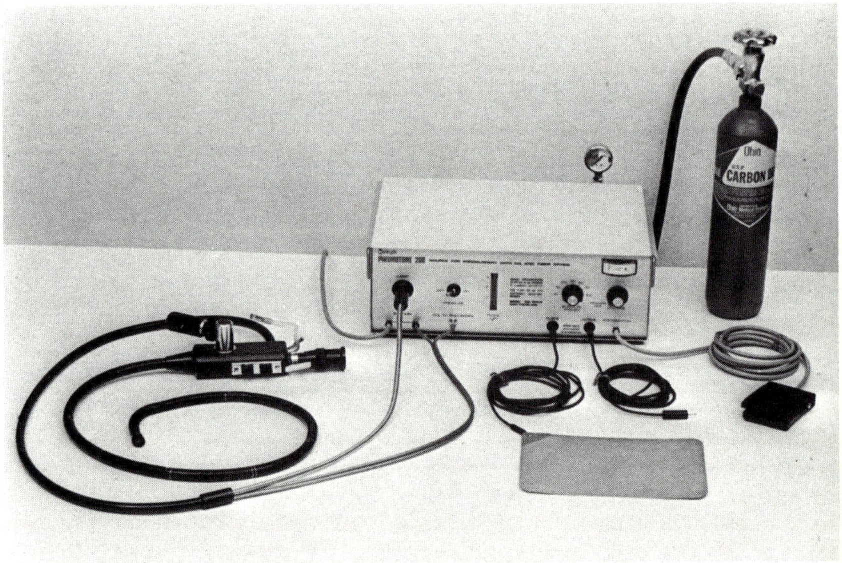

Fig. 449.
ACMI coloscope, light and electricity source, CO_2 cylinder, indifferent plate, active cable, etc.

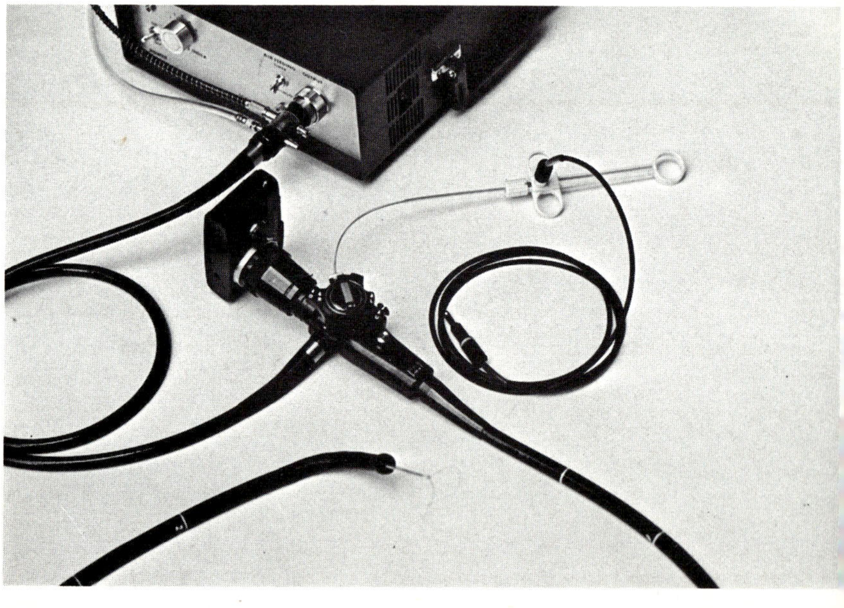

Fig. 450.
Olympus coloscope with diathermy snare, light source, live cable, etc.

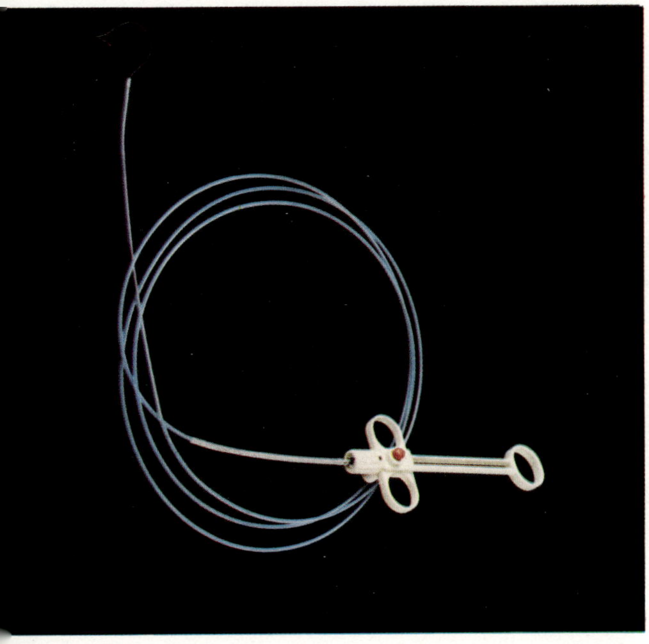

'ig. 451.
)lympus diathermy snare. This
nare is elliptical and semirigid.
)iameter 15 or 25 mm.

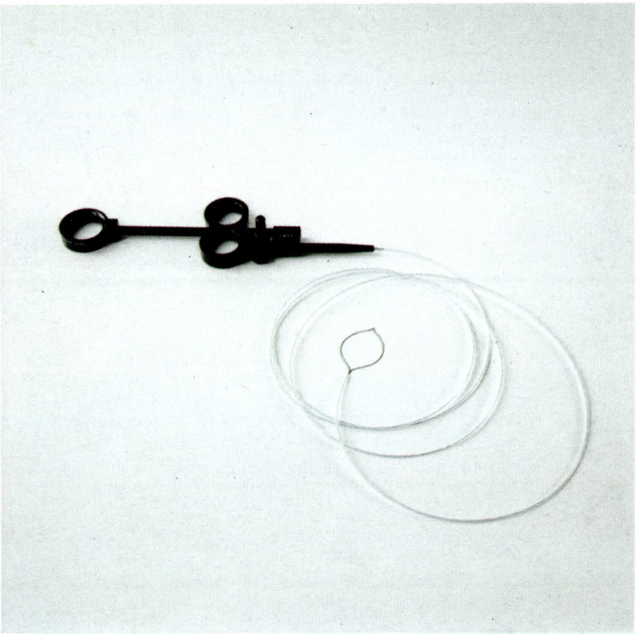

Fig. 452.
ACMI diathermy snare. This snare
is round-shaped, flexible, and
measures about 30 mm in diameter.

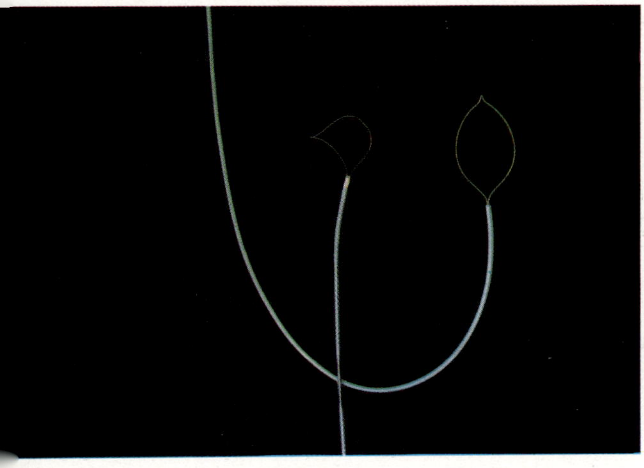

'ig. 453.
)lympus and ACMI snares showing
he different shape of the loops.

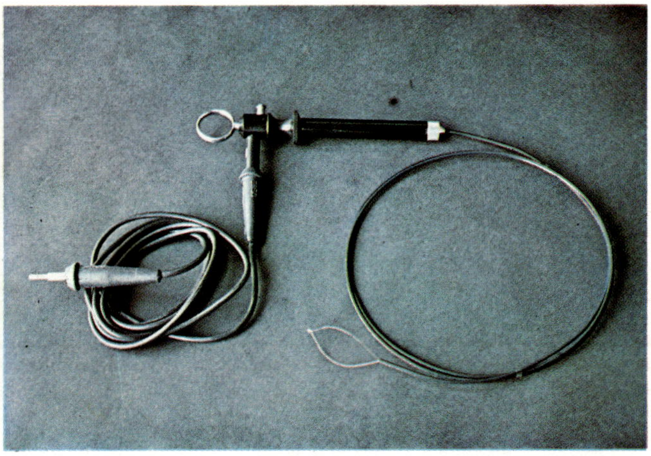

Fig. 454.
Storz diathermy snare. This snare is
round-shaped, semirigid, and
measures about 25 mm in diameter.

)iathermy snares

\ scissors-tip electrosurgical forceps has been described.
t consists of a length of wire contained in an insulating sheath
ınd provided with a scissors-like double-bladed tip.
'his forceps can only be used to remove stalked polyps.
\ diathermy/biopsy forceps has recently been introduced with
vhich small polyps can be ablated endoscopically.
t consists of a spiked biopsy forceps with an insulating Teflon
heath with which small lesions can be excised and
imultaneously electrocoagulated.

technique

Polypectomy can only be performed after the bowel has been completely cleared of faeces. Small amounts of water do not constitute a contraindication.
The diathermy snare is inserted through the biopsy channel of the fibrecoloscope. It is then looped over the polyp and closed around the stalk. Care must be taken not to close the snare too tightly to avoid cutting the stalk. The snare is withdrawn almost completely into the sheath. The snared polyp will be seen to swell and become cyanotic. Little to-and-from and lateral movement of the coloscope tip can be used to check the correct position of the snare.
An inert gas such as CO_2 or nitrogen is insufflated into the bowel lumen before applying electric current. At this point the earthing of all the equipment is completed and checked.
The polyp is then moved away from the opposite wall of the bowel and transected by applying coagulating high frequency current of appropriate potency. High frequency (130 kHz - 500 kHz) generators of 40 W - 100 W maximum potency are usually used.
When operating on a pedunculated polyp the snare must not be applied at the base of the stalk touching the bowel wall but about 1 cm up the stalk to avoid the risk of burning and perforation (Figs. 455 & 456). Any residual part of the stalk can be removed later (Fig. 457).
Sessile polyps require a slightly different procedure. The diathermy snare is looped around the base of the polyp on the bowel wall and is then withdrawn and closed. Before applying electric current the polyp is pulled towards the bowel lumen so as to create an artificial stalk.
This reduces the burnt area and minimizes the risk of perforation (Fig. 458). Some prefer to inject saline solution at the base of the polyp to form a protective layer.
Large lobulated polyps which cannot be removed in one piece are removed piecemeal. The proximal lobulation is excised first and the remaining parts of the polyp are then transected from the base (Figs. 459 & 460). Should the view be obscured by bleeding the operation must be postponed until the next day.

Fig. 455.
The diathermy snare is looped over the polyp and is then closed about 1 cm from the base of the stalk.

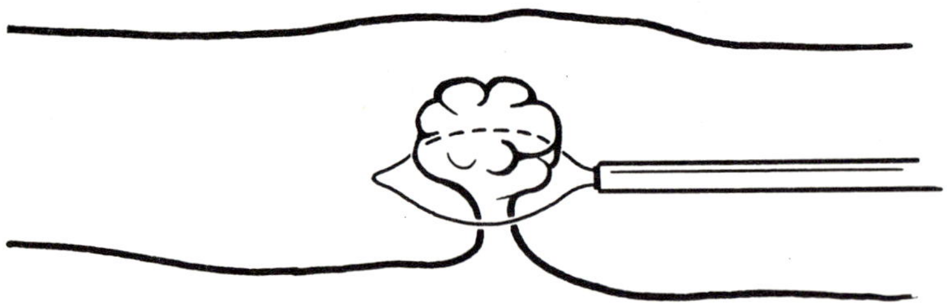

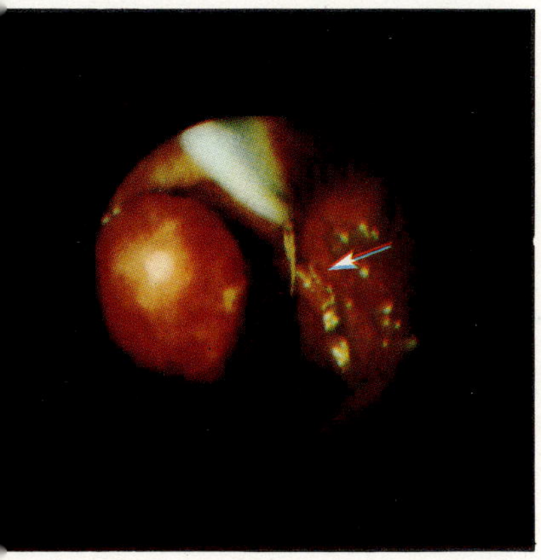

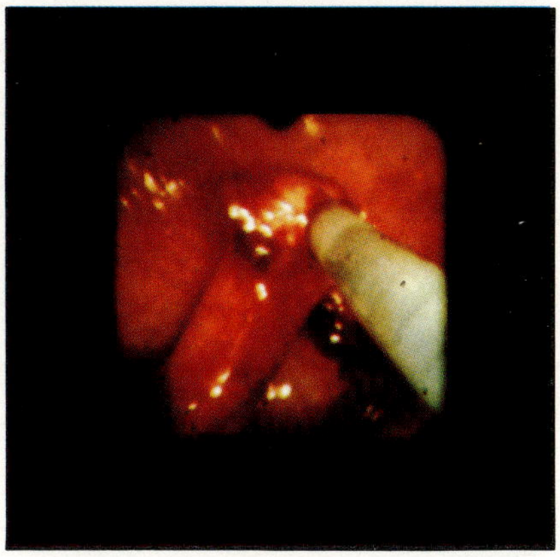

Fig. 456.
Endoscopic view. The partly withdrawn metal snare begins to close around the stalk about 1 cm from its base.

Fig. 457.
The diathermy snare has removed the body of the polyp. There remains a long stalk which will be removed later in the course of the same session.

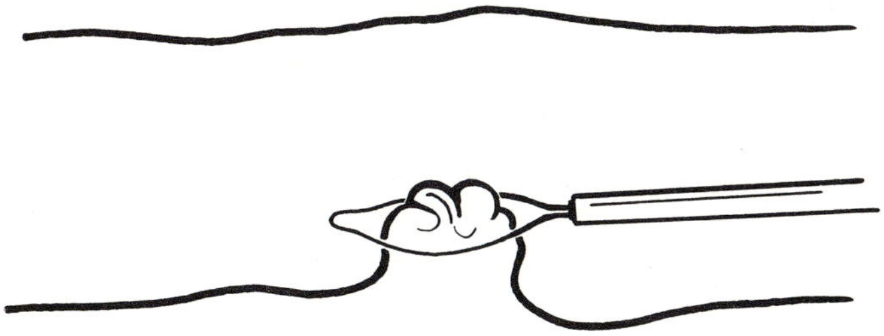

ig. 458.

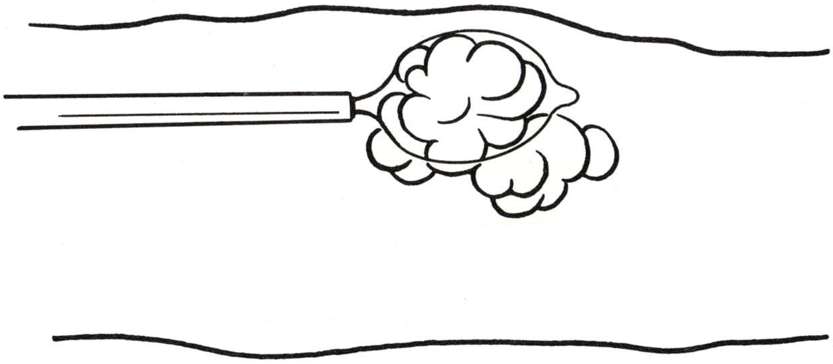

ig. 459.

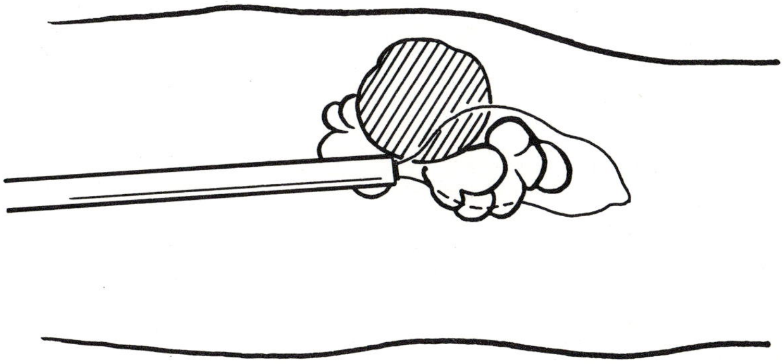

ig. 460.

retrieval of the excised polyp

It is important that the transected polyp be retrieved for histologic examination. This can be achieved:

1) by aspiration onto the tip of the coloscope;
2) by removing it with the aid of the biopsy forceps or special retrieving forceps or basket or of the snare itself;
3) by recovering it in the stools.

Aspiration onto the tip of the coloscope:
this is the simplest method when there is only one polyp. After transection the diathermy snare which obstructs the suction channel is withdrawn rapidly and the site of the operation is inspected for bleeding. The excised polyp is located and the tip of the coloscope is advanced until it touches it. The polyp is then aspirated against the tip of the coloscope.
This method has certain drawbacks:

1) the mucosa of the colon cannot be inspected during withdrawal of the coloscope;
2) forcible aspiration of the polyp onto the tip of the coloscope may obstruct the suction channel or, worse still, may cause obstruction at the connection of the fibrescope to the suction apparatus. This is most liable to occur in the case of small polyps and when the orifice of the suction channel of the fibrescope measures more than 2 mm in diameter.

Retrieval of excised polyps using biopsy forceps, retrieval forceps or baskets or diathermy snare.
With these accessories it is possible to retrieve the polyp and at the same time to inspect the mucosa of the colon as the fibrescope is withdrawn. However, one cannot always be sure that the whole polyp will be recovered because it may break up in the jaws of the forceps or drop off during withdrawal of the fibrescope.
Should the foregoing methods prove ineffective or difficulties arise, the polyp can be recovered by giving a cold water enema. This is the best procedure in multiple polypectomies because it makes it possible to recover all the excised growths.
The stiffening tube can also be used to advantage. The tube is kept in place while the coloscope is withdrawn to retrieve the polyp attached to its tip or clamped by a forceps.
Reinsertion of the coloscope through the stiffening tube for further polypectomies or to inspect the site of operation is thus very easy.

'ig. 461.
ACMI operative coloscope featuring
wo biopsy channels which make
ossible the simultaneous
ransection and retrieval of the polyp.

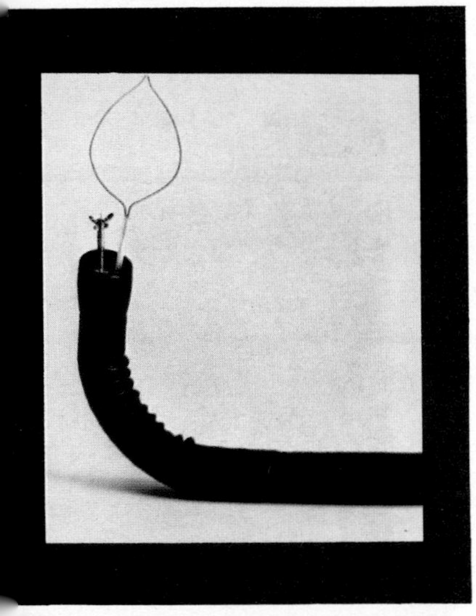

Fig. 462.
ACMI coloscope featuring a 4 jawed retrieval forceps.

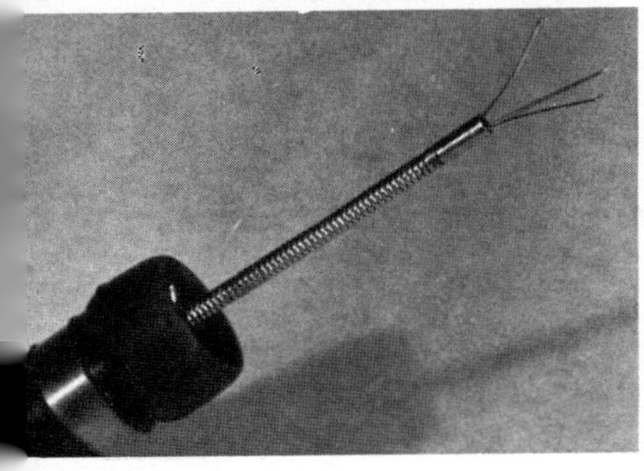

ig. 463.
)lympus coloscope featuring a
-jawed retrieval forceps.

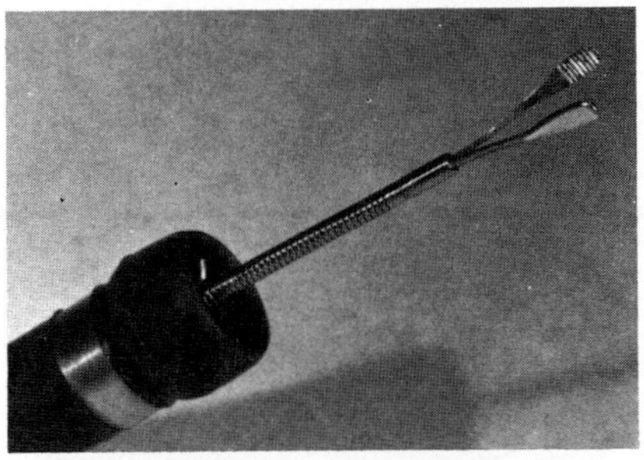

Fig. 464.
Olympus coloscope featuring a flat-bladed forceps.

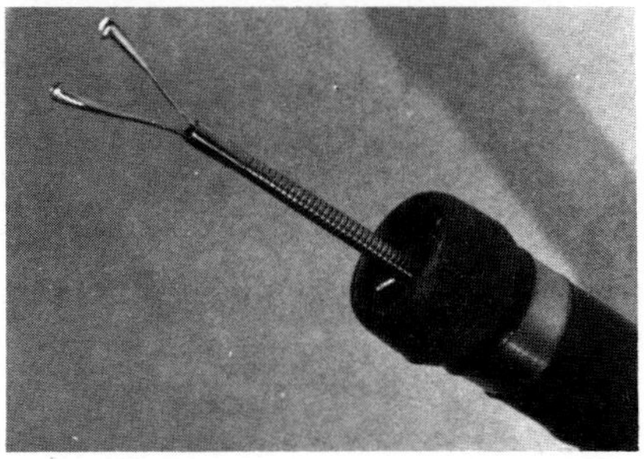

Fig. 465.
Olympus coloscope featuring a hooked forceps.

potential hazards of endoscopic electrosurgery

Endoscopic polypectomy entails certain hazards like any other type of electrosurgical operation:

1) electrocution due to faulty insulation;
2) burns caused by accidental contact with metal parts carrying high frequency electric current, either connected t the generator or due to electromagnetic induction phenomena (rings, necklaces, etc.);
3) partial or complete breakdown of an artificial cardiac pacemaker due to induction phenomena by high frequency electric current.

technical accidents

Breakage of the snare or twisting of the snare to form an 8 loop, especially when operating on large or pseudolobulated polyps.

precautions

To avoid the risks connected with endoscopic electrosurgery the endoscopist must check that:

1) every piece of equipment is supplied with the correct voltage;
2) the cables are all effectively insulated;
3) every piece of equipment is effectively earthed;
4) the surface of the indifferent plate is wide and even;
5) all connections are correctly made and the by-pass cable connects the endoscope with the indifferent plate. This connection will earth any dispersed high frequency current
6) the power supply is adequate; power should not be excessive (20 to 40 Watts) and should be regulated according to the expected surgical effect;
7) the polyp must not touch the opposite wall of the bowel because this can cause burning or perforation on account o the high density of the electric current when the point of contact is small. If the polyp touches the opposite wall this should preferably occur over a large area;
8) an inert gas must be insufflated (CO_2, nitrogen) to avoid the risk of explosion due to the presence of inflammable gases in the bowel lumen.

limitations

1) presence of diverticula close to the polyp;
2) large sessile growths infiltrating the bowel wall;
3) presence of dilated veins close to the polyp.

Complications

Bleeding

Bleeding is an uncommon complication (0.2%-0.3% of cases). The most common cause of bleeding is defective electrocoagulation, as may be the case in the presence of a broad stalk. More rarely it may be due to the untimely fall of the scab. In this case treatment should consist in cauterization of the bleeding vessel by means of a thin-pointed electrode. If bleeding is slight, it can usually be arrested by sprinkling a few drops of epinephrine onto the bleeding area through a Teflon tube. The slight bleeding due to injury by the diathermy snare cannot be considered a complication.

Perforation

Perforation is due to the potent diathermic effect of high frequency electric current, the most common cause being transection of the polyp too close to the bowel wall. This complications is also very uncommon (0.5‰ of cases).

Fig. 466.
Barium enema showing a roundish filling defect in the upper part of the descending colon with an apparently broad base on the lateral wall, attributable to a polypoid growth.

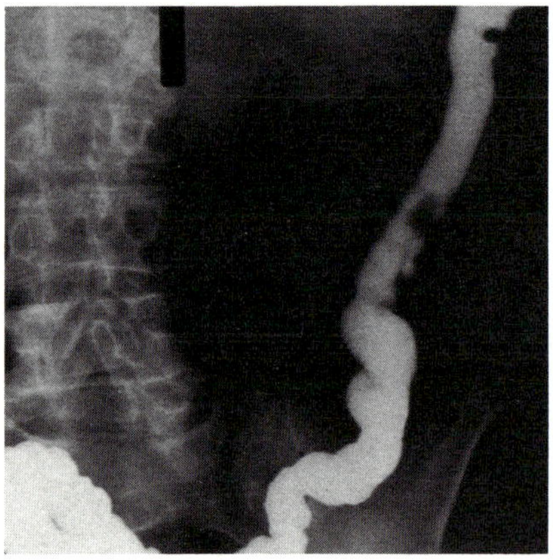

Fig. 467.
The diathermy snare has closed over the polyp (which was detected in the middle third of the descending colon) about 1 cm from its base. The lesion has become slightly swollen and cyanotic.

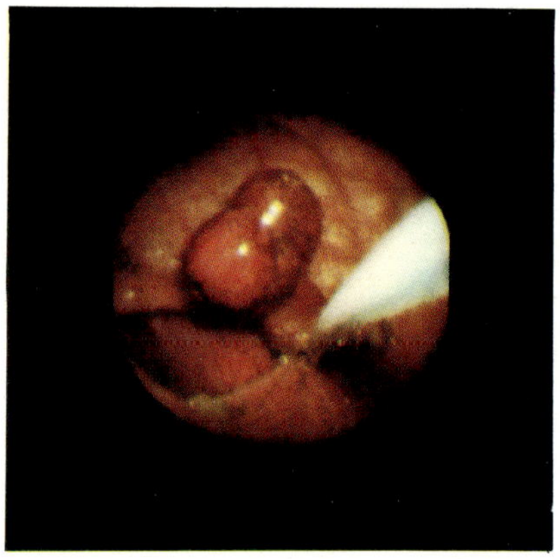

Fig. 468.
Area from which the polyp was excised. Bleeding is slight. Immediately after the operation the surrounding mucosa has an oedematous and swollen appearance

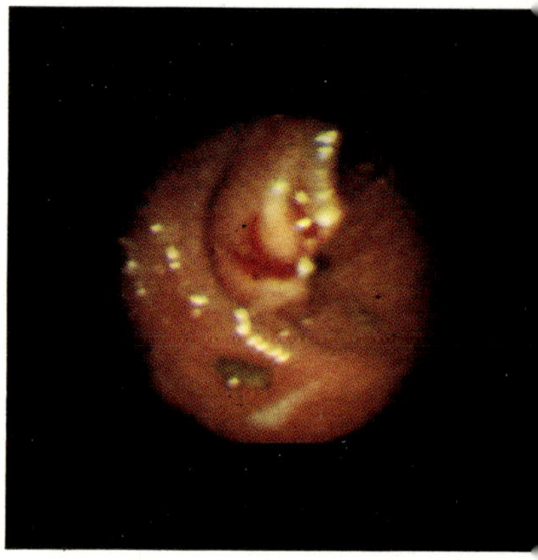

Fig. 469.
The same area 10 minutes after polypectomy. The mucosa is flat.

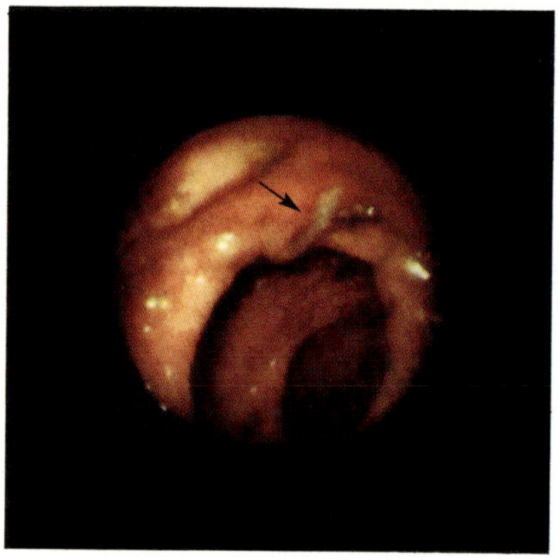

Fig. 470.
The excised polyp is in the bowel lumen.

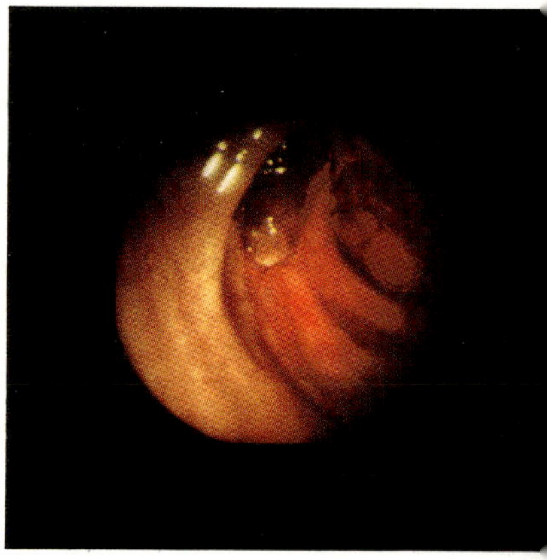

Fig. 471.
Polyp retrieved by aspiration onto the tip of the coloscope. The site of electrosurgical transection is clearly visible on the stalk.

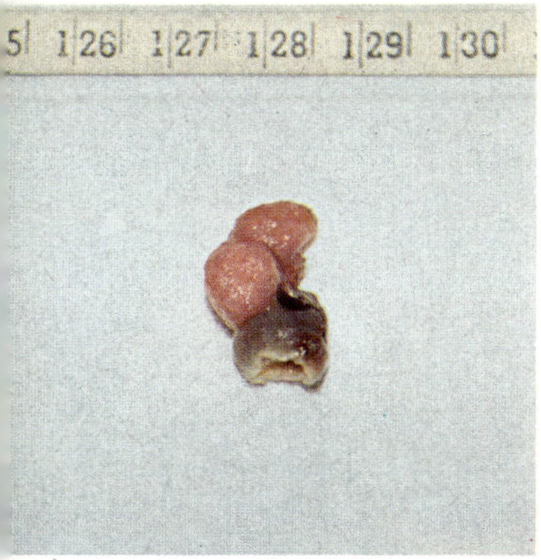

Fig. 472.
Histologic aspect of the excised polyp (adenomatous polyp).

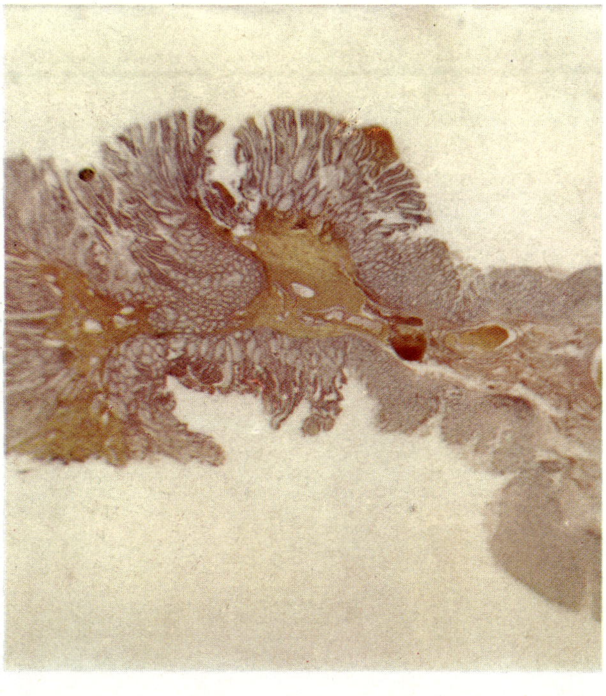

Fig. 473.
Sessile polyp of the sigmoid-descending colon junction. The polyp has a slightly irregular surface and shallow ulcerations.

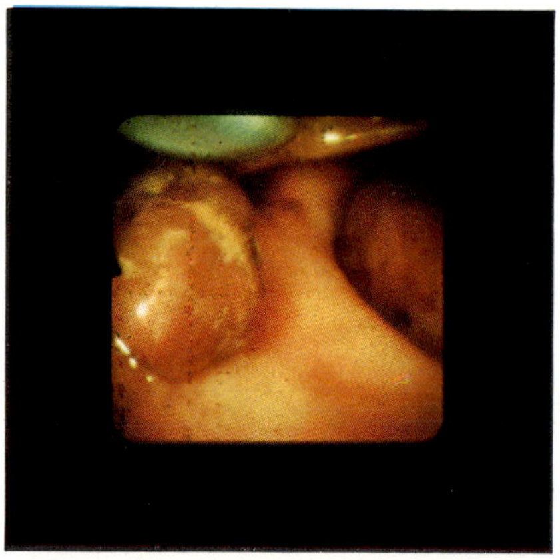

Fig. 474.
A semirigid elliptical diathermy snare is placed in front of and a little above the growth.

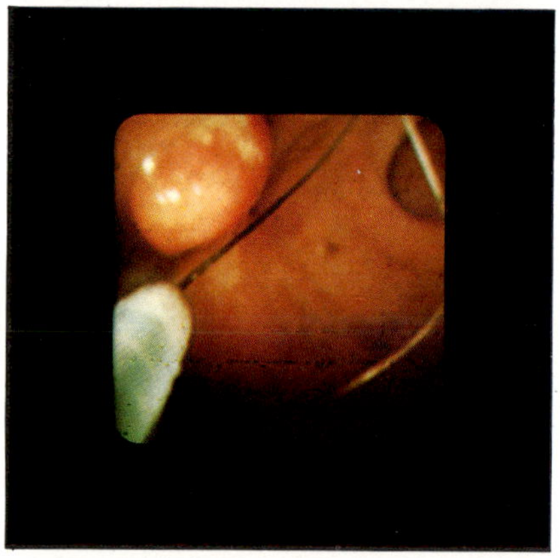

Fig. 475.
Site of electrical transection and coagulation.

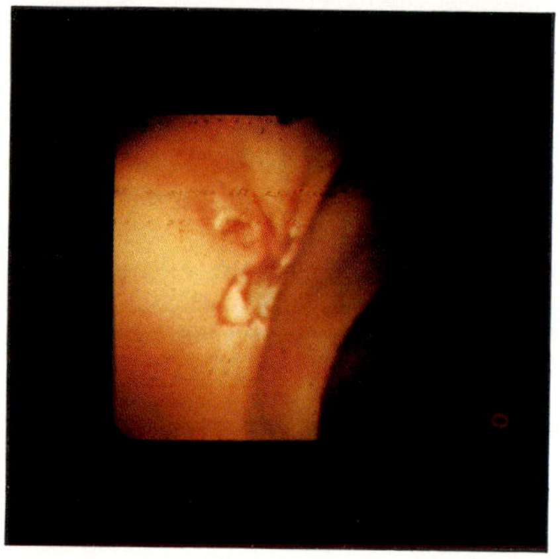

Fig. 476.
Polyp retrieved by means of the biopsy forceps.

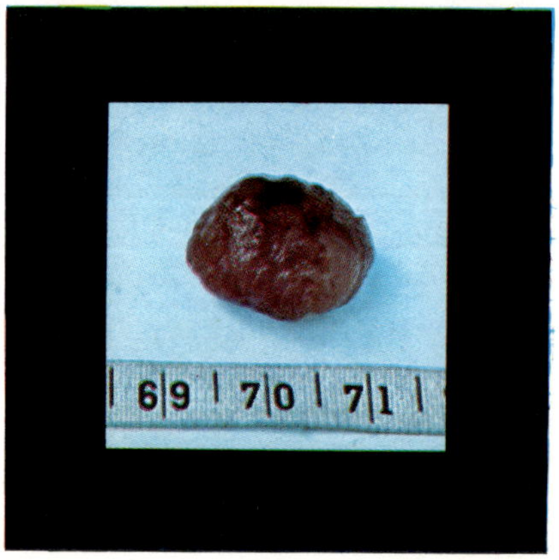

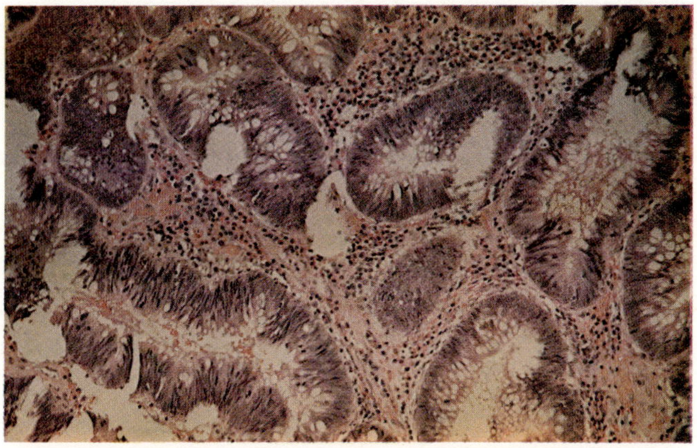

Fig. 477.
Histologic appearance of a specimen obtained with biopsy forceps (adenomatous polyp with atypical features).

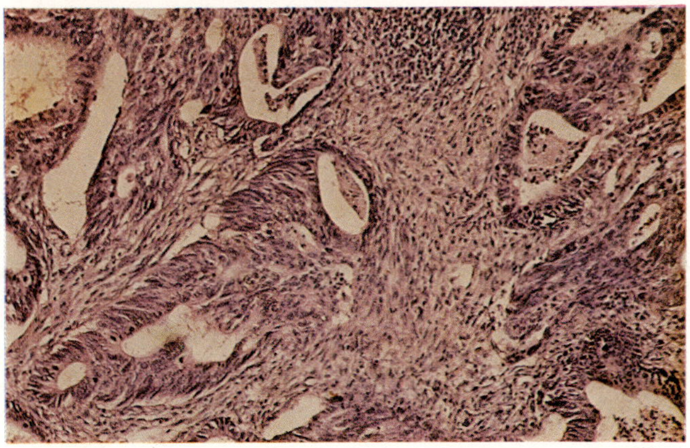

Fig. 478.
Histologic appearance of the excised lesion (adenomatous polyp: detail showing in situ carcinoma).

Fig. 479.
Diathermy snare looped around a long-stalked polyp of the colon.

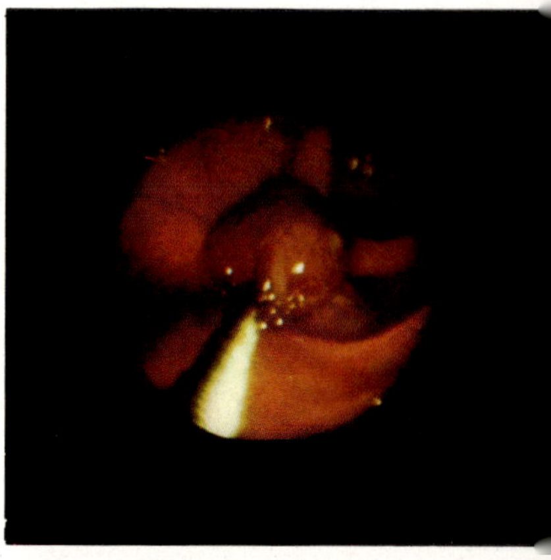

Fig. 480.
Snare closed around the body of the polyp.

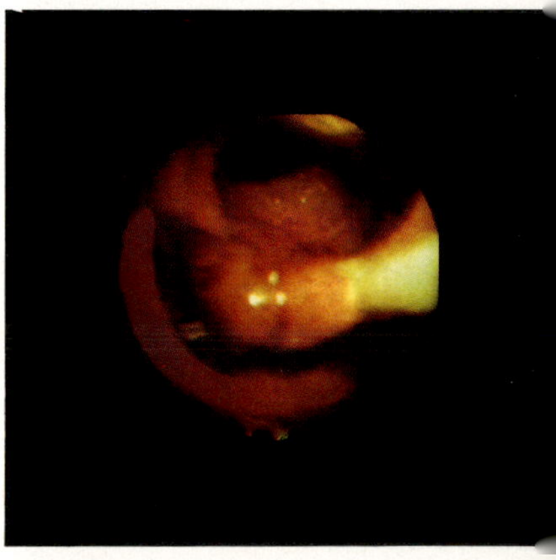

Fig. 481.
The body of the polyp, separated from the stalk, has fallen into the bowel lumen.

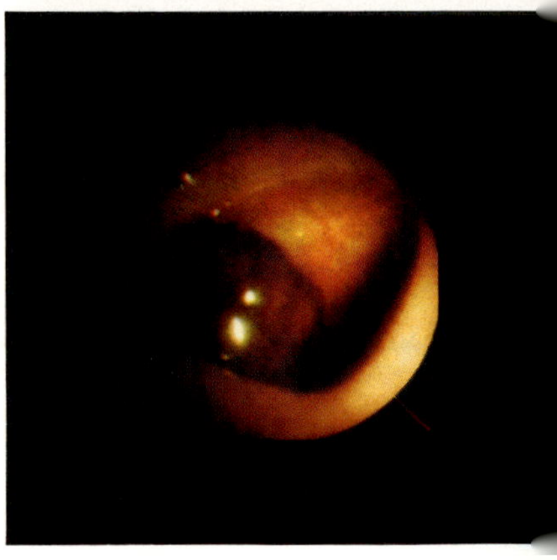

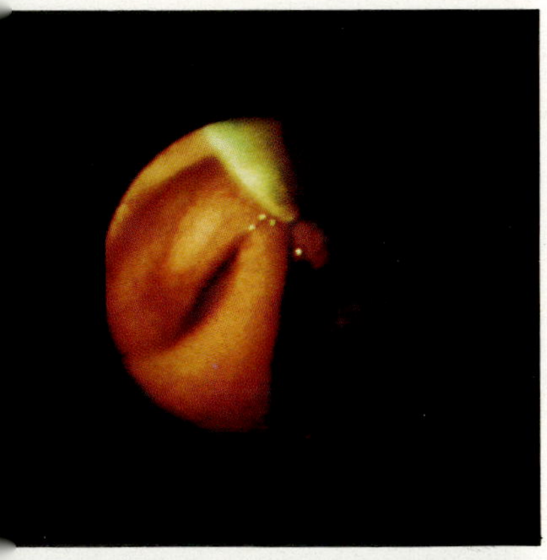

Fig. 482.
The residual long stalk is clamped by the snare and removed in one piece from its base.

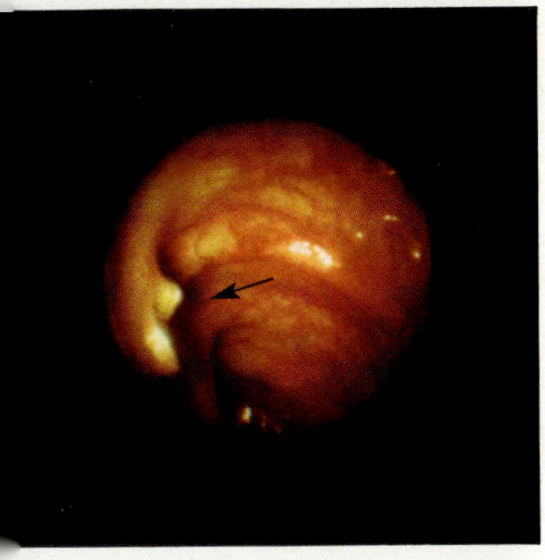

Fig. 483.
Site of electrosurgical transection of the residual stalk.

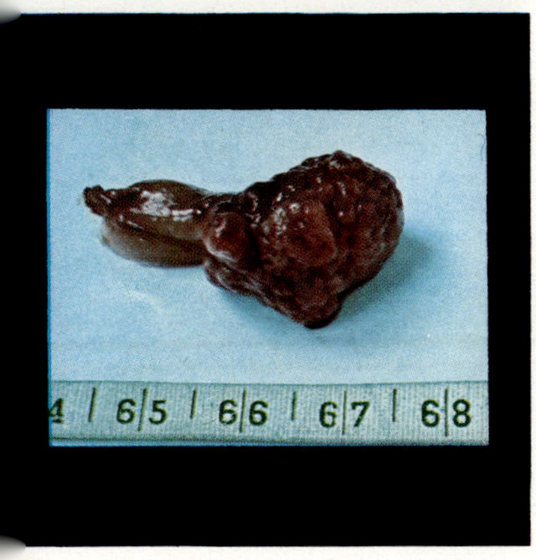

Fig. 484.
The excised polyp.
Histologic diagnosis:
adenomatous polyp.

Fig. 485.
Small sessile growth on the wall of the sigmoid colon.

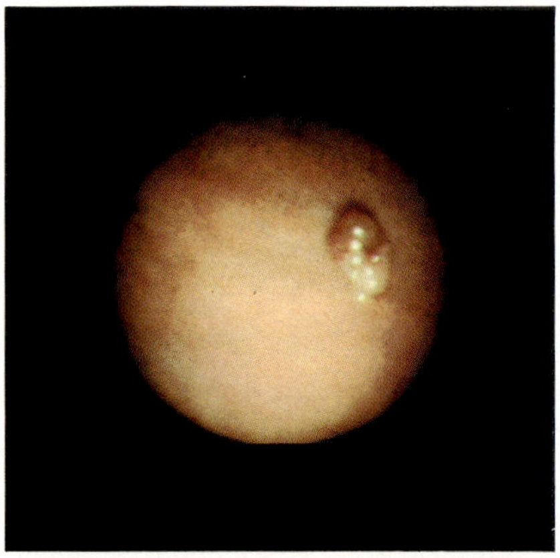

Fig. 486.
Ulcerated area at the site of polypectomy.

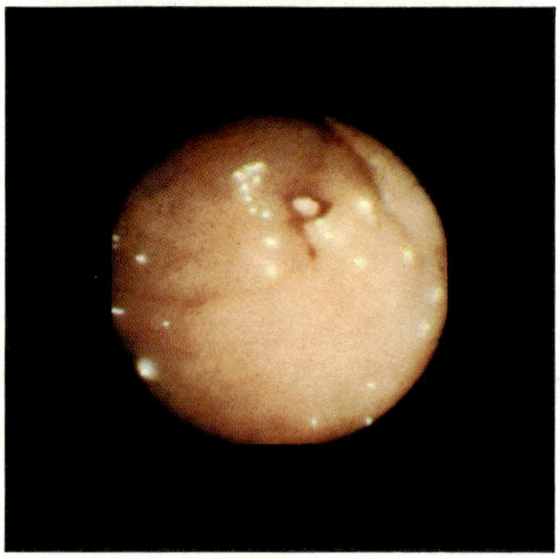

Fig. 487.
Excised polyp retrieved with the aid of a retrieving basket.

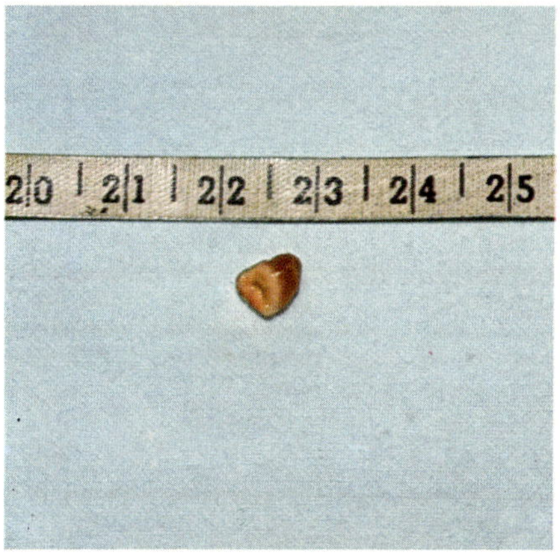

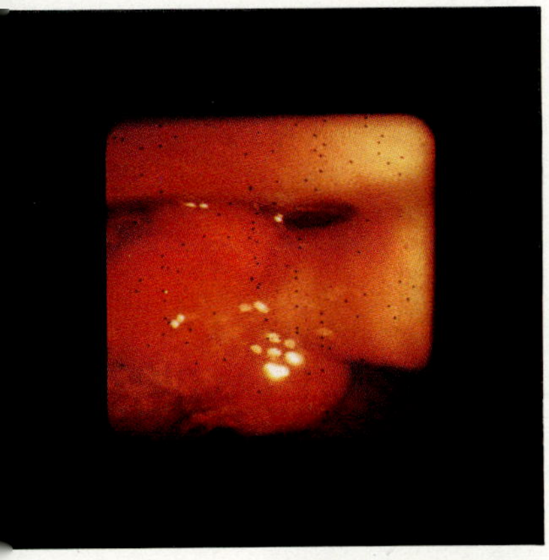

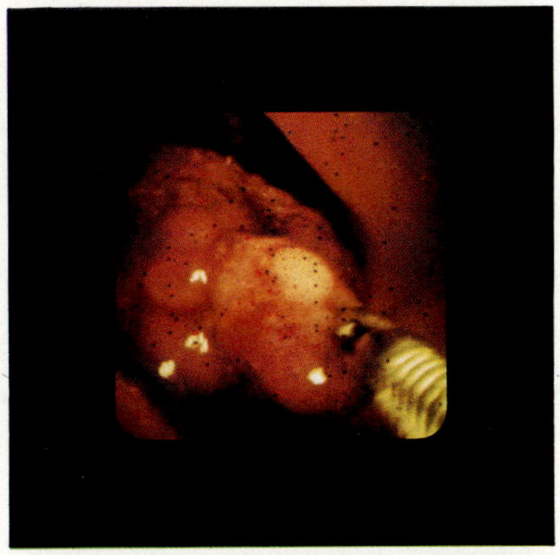

Fig. 488.
Large pseudolobulated polyp with broad, short stalk.

Fig. 489.
The excised polyp is retrieved with the aid of the biopsy forceps.

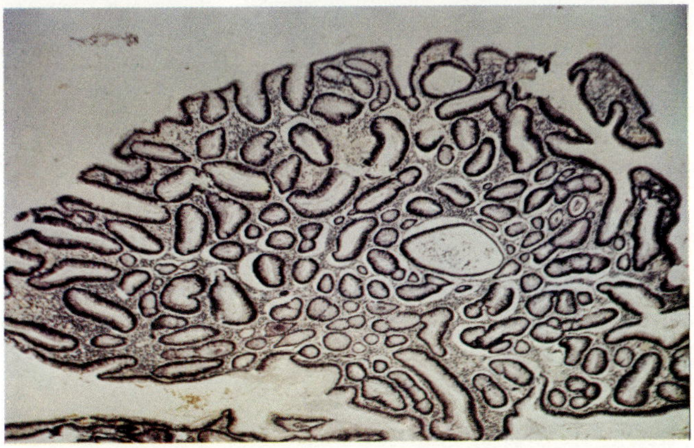

Fig. 490.
Histologic appearance of the excised polyp (adenomatous polyp with atypical glandular structure).

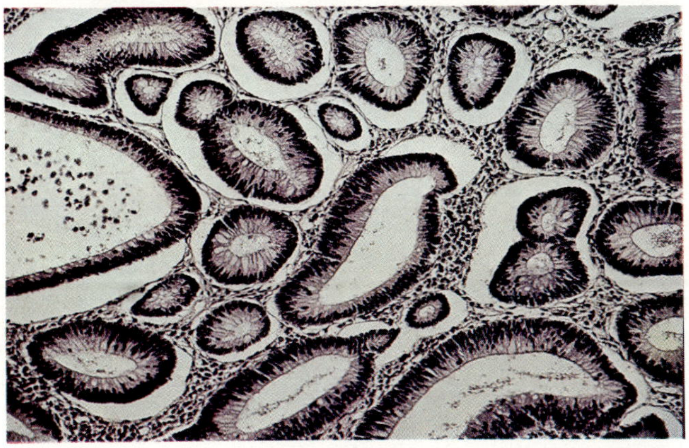

Fig. 491.
Detail of Fig. 490 with high magnification.

Fig. 492.
24 hours after coloscopic polypectomy the patient had severe rectal bleeding for which she was immediately operated on. Surgical specimen showing the residual stalk after polypectomy.

Part four

References

ARMENGOL-MIRO J. R.	Valor de la colonoscopia en el diagnostico de las enfermedades del intestino grueso. Comunicazion al Congreso International de Cirugia - Barcelona 1973.
ARMENGOL-MIRO J. R.	Experiencia personal de fibrocolonoscopia. Medicina Clinica 1974.
ARULLANI A., PAOLUZZI P., CAPURSO L.	Endoscopy of the colon. Proc. Ist. Europ. Congr. Digestive Endoscopy, pp. 2-6, Prague 1968.
AXON A. T. R., COTTON P. B., PHILLIPS I., AVERY S. A.	Disinfection of Gastrointestinal fibre endoscopes. Lancet, pagg. 656-658, april 13, 1974.
BACON H. E., HOLOMAN M. B., CHINPRAHAST K.	Pancolonoscopy: is it a valuable procedure? Dis. Colon Rect., 6, 311, 1963.
BANCHE M., ROSSINI F. P.	La polipectomia endoscopica dello stomaco e del colon. Congresso Società Endoscopia Digestiva Italiana, Venezia 1973.
BANCHE M., ROSSINI F. P., BONARDI L.	The colonoscopy in the relatively early diagnosis of colon tumors. Second International Symposium on Cancer Detection and Prevention—Bologna 1973.
BANCHE M., ROSSINI F. P., FERRARI A., CROCE G.	Le rôle de la polypectomie endoscopique dans le diagnostic, la prophylaxie et la therapie du cancer du tube digestif. XI Congreso SMIER—Madrid, 1974.
BANCHE M., ROSSINI F. P., FERRARI A., ROATTA L., GILLI E., CIRILLO R.	The role of coloscopy in the differential diagnosis between idiopathic ulcerative colitis and Crohn's disease of the colon. 39th Annual Convention—American College of Gastroenterology. Florida october 1974.
BANEZ A. V.	Operative fiberoptic colonoscopy. Pers. Comm. 1973.
BARTELHEIMER W., REMMELE W., OTTENIANN R.	Coloscopic recognition of hemangiomas in the colon ascendens. Case report of so-called cryptogenic gastrointestinal bleeding. Endoscopy, 4; 100-114, 1972.
BECKER H., SCHUELKE K.	Koloskopie—Ein Verfahrren zur intraoperativen endoskopischen tumordiagnostik des dickdarms. Endoscopy, 3: 182-184; 1970.
BELDA A., BERENGUER J., RODRIGO M., GARRIDO G., PERTEJO V., LAGUENA J.	Fibrocolonoscopie: Experience personnelle. Comm. 2e Congres Europ. Endosc. Digest., Paris 1972.

BENNETT J. A., SALMON P. R., BRANCH R. A., BASKETT P. J. F., READ A. E. A.	The use of inhalational anaesthesia during fibre-optic colonoscopy. Anaesthesia, 26, 294-297 1971.
BERCI G., MORGENSTERN L.	Diagnostic colonoscopy and colonoscopic polypectomy. Arch. Surg. 106: 818-819, 1973.
BERCI G., PANISH J. F., SCHAPIRO M., CORLIN R.	Complications of colonoscopy and polypectomy. Gastroenterology 67: 584-585, 1974.
BEREZOV Y. E., SOTNIKOV V. N., KORNILOV Y. M.	Colonoscopy in diagnosing disease of the large intestine (Russian). Vestn. Akad. med. Nauk., 27(2), 65-69; 1972.
BERRY LEONIDAS H.	Gastrointestinal pan-endoscopy. Charles & C. Thomas—Publisher—1974 Illinois (USA).
BLANKENHORN D H., HIRSCH J., AHRENS E. H. jr.	Transintestinal intubation : technic for measurements of gut length and physiologic sampling at known loci. Proc. Soc. Exp. Biol. (N.y.), 88: 356; 1955
BLOOM B. S., GOLDHABER S. Z., SUGARBAKER P. H., O'CONNOR N. E.	Fiberoptics: morbidity and cost. New. Engl. J. Med., 288, 368-369; 1973.
BOCKUS H. L.	Gastroenterology, Vol. II, Saunders, Philadelphia.
BOGNEL J. C., LIGUORY A., BITOUN A., DUHAMEL J.	L'exerese des polypes par coloscopie. Vie Medicale 36, 4615-4626—Novembre 1973/3.
BOLT R. J.	Sigmoidoscopy in detection and diagnosis in the asymptomatic individual. Cancer (Philad.), 28, 121: 1971.
BOZZINI Ph., SCHINDLER R.	Gastroscopy. U. of Chicago Press, 1937.
CADRANEL S., RODESCH P., CREMER M.	Technique et premiers resultats de la colonoscopie chez l'enfant. Comm. 2e Congres Europ. Endosc. Digest., Paris 1972.
CAVENACH A. J. M., RINTOUL R. F.	Cancer of the bowel. Brit. Med. J., 3: 165-166; 1972.
CHABANON R.	La colonoscopie. 19e Journées Lyonnaises de Medicine, 1972. Rev. Lyon. Med. (Suppl.), 19-20; 1972.
CHABANON R.	La polypectomie coloscopique. Acta gastro-ent. Belg., vol. 37, 166-171, 1974.
CHABANON R. ARNOUS J.	La coloscopie dans la rectocolite hemorragique. Arch. Mal. Appar. Dig. 61, 369-372.

CLASSEN M.	Progress report. Fiberendoscopy of the intestines. Gut 12: 330; 1971.
CLASSEN M., FRUHMORGEN P., KOCH H., DEMLING L.	Peroral enteroscopy of the small and the large intestine. Endoscopy, 4: 157-162; 1972.
CLASSEN M., KOCH H., FRUHMORGEN P.	Perorale enteroskopie. Med. Trib., 47a, 3; 1971.
COLCHER H.	Colonoscopie, quo vadis. Ann. Gastroenterol. Hepat. 7: 247-250; 1971.
COLCHER H.	Progress in gastrointestinal endoscopic instrumentation in the past decade. Gastrointest. Endosc. 17: 169-172, 1971.
CORNET A., BARBIER J. P., CARAYON P., MARCHE C.	Polypes coliques. Interet du depistage et de l'identification par la coloscopie. Arch. Franc. Mal. Appar. Dig., 60: 261; 1971.
COTTON P. B., WILLIAMS C. B.	Fibreoptic instruments for gastrointestinal endoscopy. Brit. J. Hosp. Med., 8, 35-44 (Equipment Suppl., Nov.); 1972.
CRESPON B., HOUSSET P., CAMPORA A., PAOLAGGI J. A., DEBRAY Ch.	Apport de la coloscopie d'urgence dans les hemorragies digestives basses. Acta Endoscopica et Radiocinematographica Tome III - n. 5 - pagg. 133-134 - novembre 1973.
CRESPON B., HOUSSET P., DEBRAY Ch.	La coloscopie dans les colites organiques chroniques. Ann. Gastroentérol. Hépat., 9, 529-535, 1973.
CURTISS L. E.	High frequency currents in endoscopy: a review of principles and precautions. Gastrointestinal Endoscopy; 1973.
DEAN A. C. B.	Problems with coloscopic polypectomy. Endoscopy 5: 121-123; 1973.
DEAN A. C. B.	Coloscopy in diagnostic problems in the large bowel. Acta gastro-ent. belg., Vol. 37, 145-149, 1974.
DEAN A. C. B., NEWELL J. P.	Colonoscopy in the differential diagnosis of carcinoma from diverticulitis of the sigmoid colon. Brit. J. Surg., 1973.
DEAN A. C. B., SHEARMAN D. J. C.	Clinical evaluation of a new fiberoptic colonoscope. Lancet, 1, 550-552; 1970.
DEAN A. C. B. SHEARMAN D. J. C.	A clinical evaluation of the Olympus cf-sb fibreoptic colonoscope. Proc. IInd World Congress of Gastrointestinal Endoscopy. Rome—Copenhagen 1970.
DEDDISH M. R., FAIRWEATHER W. H.	Colotomy and coloscopy. Cancer, 6, 1021; 1953.

DEDDISH M. R., HERTZ R. E.	Coloscopy in the treatment of mucosal polyps of the colon. Surg. Clin. North America, 37, 1287; 1957.
DEDDISH M. R., HERTZ R. E.	Colotomy and coloscopy in the management of neoplasm of the colon. Dis. Colon Rect., 2, 133; 1959.
DEDDISH M. R., HERTZ R. E.	Colotomy and coloscopy in the management of mucosal polyps and cancer of the colon. Amer. J. Surg., 90, 846-849; 1955.
DE LA SANTA LOPEZ J. et al.	Fibrocolonoscopia: téchnica y valoracion clinica. Rev. Espan. Enf. Apar. Dis. 39: 651-660, 1973.
DESORMEAUX A. J., SCHINDLER R.	Gastroscopy. U. of Chicago Press, 1937.
DEYHLE P.	Symposion - Duodenoskopie, koloskopie. I. Eeco-kongress, Munchen; 1970.
DEYHLE P.	Technik und klinische bedeutung. Endoskopiekongress, Dtsch. Ges. G. Endoskopie, Erlangen 1971.
DEYHLE P.	Flexible steel wire for the maintenance of the straightening of the sigmoid and transverse colon during coloscopy. Endoscopy, 4: 36-38, 1972.
DEYHLE P.	A new technique for the fiberendoscopic passage through the s-type loop of the sigmoid colon. Endoscopy, 4; 102-104; 1972.
DEYHLE P.	A plastic tube for the maintenance of the straightening of the sigmoid colon during colonoscopy. Endoscopy, 4, 224-226; 1972.
DEYHLE P., DEMLING L.	Coloscopy-Technique; Results, indication. Endoscopy, 3: 143-151; 1971.
DEYHLE P., JENNY S., FUMAGALLI I.	Endoskopische polypectomie in proximalen kolon. Dtsch. Med. Wschr., 98: 219; 1973.
DEYHLE P., JENNY S., FUMAGALLI I., LINDER E., AMMANN R.	Endoscopy of the whole small intestine. Endoscopy, 4: 155-157; 1972.
DEYHLE P., LAGIARDER F., JENNY S., FUMAGALLI I.	A method for endoscopic electroresection of sessile colonic polyps. Endoscopy, 5: 38-40; 1973.
DEYHLE P., OTTENJANN R.	Zur transintestinalen intubation. In: R. Ottenjann (ed.), Fortschritte der Endoskopie, BD. 2. Schattauer, Stuttgart 1970.
DEYHLE P., PAUL F.	Die endoskopie des proximalen dickdarmes. Therapiewoche, 20: 1803; 1970

DEYHLE P., PAUL F., OTTENJANN R.	Die proximale coloskopie. In "Aktuelle Gastroenterologie 69" Karger Verlag, Basel 1970.
DEYHLE P., SEUBERTH K., JENNY S., DEMLING L.	Reports on new instruments and new methods. Endoscopic polypectomy in the proximal colon. Endoscopy, 2: 103-105; 1971.
DILAWARI J. B., PARKINSON C., RIDDEL R. H., LOOSE H., WILLIAMS C. N.	Colonoscopy in the investigation of ulcerative colitis. (Abstr.) Gut, 14, 426; 1973.
DIVE C., VANHEUVERZWIJN R.	Aspects endoscopiques de la colite granulomateuse. Acta gastro-ent. belg., vol. 37, 150-153; 1974.
DUHAMEL J., BOGNEL J. C., NGO-BINH, LIGUORY C.	Un an d'usage du colonoscope long a l'hopital Saint Antoine. Comm. 2e Congres Europ. Endosc. Digest., Paris 1972.
EDDY H. J.	Intracolonic electro-snare polypectomy using flexible fiberoptic colonoscopes. Comm. 2e Congres Europeen d'Endoscopie Digestive. Paris 1972.
EDDY H. J., FINNERTY U. R.	Case report: multiple undiagnosed polyps found during colonoscopy on a three-years-old child. Comm. 2e Congres Europ. Endosc. Digest., Paris 1972.
ESPINER H., SALMON P. R., TEAGUE R., READ A. E.	Operative colonoscopy. Brit. Med. J., 1, 453-454; 1973.
EWE K.	Endoskopie des rektums und kolons. Therapiewoche, 19: 567; 1969.
EWE K., WANITSCHKE R.	Coloscopy as a routine precedure. Proc. IInd World Congress of Gastrointestinal Endoscopy; Rome-Copenhagen 1970.
FAGGIOLI F., MATTEI M.	La colonscopia. Gazzetta Sanitaria, pag. 94; 1972.
FERGUSON J. A.	Total colonoscopy—A systematic surgical approach. South. Med. J. 53: 517; 1960.
FIORINI E., FRATTON A., POLETTINI F. L., GENNARI L.	Contribution to the endoscopic study of the left colon. Proc. IInd World Congress of Gastrointestinal Endoscopy, Rome-Copenhagen 1970.
FONG-MING CHANG-YOSHIHIRO SAKAI, SHINROKU ASHIRAWA	Bacterial pollution and disinfection of the colonofiberscope 1° an investigation of traditional Sterilization methods. Digestive Diseases, Vol. 18, No 11 (November 1973).
FONG-MING CHANG, YOSHIHIRO SAKAI, SHINROKU ASHIRAWA	Bacterial pollution and disinfection of the colonofiberscope II ethylene oxide gas sterilization. Digestive Diseases, Vol. 18, No 11 (Novembre 1973).

FOX J. A.	Mucosal biopsy of the colon by retrograde intubation. Results and applications. Brit. J. Surg., 54, 867; 1967.
FOX J. A.	Technik der retrograden kolonsondierung und hohen kolonbiopsie. Fortschritte der Endoskopie, BD. I. Schattauer, Stuttgart, 1969.
FOX J. A.	A fibreoptic colonoscope. Brit. Med. J., 3: 50; 1969.
FOX J. A.	Retrograde colonoscopy. Endoscopy; 4: 182-185; 1969.
FOX J. A., KREEL L.	Technique of retrograde colonic intubation and its initial application to high colonic biopsy. Gut, 8, 77; 1967.
FOX M. J. A.	Maladie de Crohn. Colonoscopie. Archives Francaises des Maladies de l'Appareil Digestif. Paris, 61 n° 4-5 p. 373-376; 1972.
FRANCILLON J., VIGNAL J., LESBROS F., PHILIPPE M., CHABANON R.	Incidences de la colonoscopie et de l'anatomie pathologique sur la tactique chirurgicale concernant les polypes du colon et du rectum. Chirurgie, 97: 322-331; 1971.
FRIENT W. G.	Colonoscopy and polypectomy. Northwest Medicine 71: 613-615; 1972.
FROTZ H., GHEORGHIU T.	Endoscopic diagnosis of colonic diseases. Leber Magen Darm. 3: 124-130, 1973.
FROTZ K., GHEORGHIU Th., PHILIPPEN R.	Koloscopie mit fiberoptischen instrumenten. Med. Welt 23: 193; 1972.
FRUHMORGEN P.	Diagnosis of inflammatory diseases of the colon by coloscopy. Acta Gastroent. belg., Vol. 37, 154-157; 1974.
GABRIELLSSON N., GRANDQUIST S., OHLSEN H.	Colonoscopy: technique, indications, and results. Lakartidningen 69/39 (4377-4382); 1972.
GABRIELLSSON N., GRANDQUIST S., OHLSEN H.	Colonoscopy with the aid of a steel wire to stiffen the fiberscope. Endoscopy, 4, 217-218; 1972.
GANGAROSA E. J., BEISEL W. R.	A simple technic for colon biopsy of the rectum. Gastroenterology, 42: 157; 1962.
GRANQUIST S., GABRIELLSSON N., OHLSEN H.	Colofiberskopie, TV genomlysning, Sam Komplement till röntgenundersöhnung. Abstracts of Medicinska rickstämman, 1971.

GREGG J. A.	Fiberoptic colonoscope as a means for introduction of pressure recording catheters into proximal colon. Clin. Found. Bull. 21/1 (30-34); 1972.
GREGG J. A. GARABEDIAN M.	Colonoscopy. Surg. Clin. N. Amer., 51: 661-665; 1971.
HANSEN L. K.	Colonoscopy: a study of 50 cases. Scand. J. Gastroent., 6, 687-691; 1971.
HARMON M. L.	Nurses' role in lower gastrointestinal endoscopy. Surg. Clin. N. Amer., 52: 1025-1029; 1972.
HAYASHI M., YUKAMA E., YUKAMA K.	Radiology compared with endoscopy for the diagnosis of the cancer of large intestine. Comm. 2e Congres Europ. Endosc. Digest., Paris 1972.
HEULLY F., GAUCHER P., LAURENT J., BAS M., BIGARD M., EANPIERRE R., MERLE M., SIMON J. M., BRICHARD J. M.	Interet de la coloscopie. Premiers resultats. Ann. Med. Nancy, 10: 1541-1548; 1971.
HIRATSUKA H.	Fibercolonoscopy in the diagnostic intestinal string derivation. Method and clinical report on ileocecal observation. IInd World Congress of Gastroint. Endoscopy, Rome-Copenhagen 1970.
IOFF H. C.	Retrograde intubation of the caecum. Amer. J. Roentgenol., 20: 226; 1928.
IRADSKY M.	Resultat och diagnosfördelning hos 81 undersökta patienter. Lakartidn 69: 4383-4387; 1972.
IRADSKY M., TOCKBRUEGGER R., VENNERHOLM M.	Erfahrungen mit der fibercolonoskopischen untersuchungsmethode bei einem material von 100 patienten. Comm. 2e Congres Europ. Endosc. Digest., Paris 1972.
ANSEN K. K.	Coloscopy: an analysis of 120 cases with special regards to the tecnique. Endoscopy (Stuttg) 5/2 (77-81); 1973.
LEINFELD G., UMP F. R.	Complications of colotomy and polypectomy. Surg. Gynec. Obstet., III, 726-728; 1960.
OBAYSHI S., MIZUNO H., ASUGAI T.	Early colonic cancer presenting as a peduncolated polyp: application of fibercolonoscopy for its detection. Gastrointestinal Endoscopy, 20: 118-120; 1974.
RATZER G. L.	Colonoscopy-current status. Dis. Colon and Rect., 7: 45-47; 1964.

KRENTZ K.	Koloskopie. I. Gemeinsame Arbeitstagung von Rontgenologen, Nuklearmedizinern und Endoskopikern. Erlangen 1971.
KULD HANSEN L.	Kolonoskopi og diodenoskopi. Ugerskr. Laeg., 132, 1477; 1970.
KULD HANSEN L.	Colonoscopy. A study of 50 cases. Scand. J. Gastroent., 6: 687-691; 1971.
KULD HANSEN L.	Colonoscopy. p. 282-286 in Gastrointestinal Hormones and Other Subjects. Fifth Scandinavian Conference on Gastroenterology, Aalbord, Denmark 1971.
KULD HANSEN L.	Colonoscopy. An examination of the diagnostical efficiency of the long colonoscope. Comm. 2e Congres Europ. Endosc. Digest., Paris 1972.
LAMBERT R., MOULINIER B.	La gastro-enterologie a l'ere de l'endoscopie. Presse Med., 79: 939-940; 1971.
LEVINSON J. D., WALL A. J., KIRSNER J. B.	The problem of carcicoma in inflammatory disease of the bowel: selective cases experiences. Southern. Med. J., 65; 209-214; 1972.
LEZAK M. B., GOLDHAMMER M.	Retroperitoneal emphysema after colonoscopy. Gastroenterology 66: 118-120; 1974.
LUX G., LUX E.	Koloskopische Polypectomie, Karzinomprophylaxe? Symposium in Erlangen-Marz 1973.
MADIGAN M. R., HALLS J. M.	The extent of sigmoidoscopy shown on radiographs with special reference to the rectosigmoid junction. Gut, 9, 355-362; 1968.
MARCOZZI G., MONTORI A.	Utilità dell'endoscopia digestiva peroperatoria nelle sindromi di difficile diagnosi. IV Congresso della Società di Endoscopia Digestiva, Siena, Nov. 1971.
MARCOZZI G., MONTORI A.	Endoscopy during operation: new diagnostic possibilities preliminary report. Chir. Gastroent. 6, 24; 1972.
MARCOZZI G., MONTORI A., CRESPI M.	Intraoperative endoscopy. Urgent endoscopy of digestive and abdominal diseases. Int. Symp., Prague Carlsbad; 1971.
MARKS G.	Electrosurgical scissor-type polyp forceps for the flexible fiberoptic colonoscope. A new instrument. Preliminary report. Submitted for pubblication.
MARKS G., MOSES M. L.	The clinical application of flexible fiberoptic colonoscopy. Surg. Clin. N. Amer 53/3; 1973.

MATHIS P., SEGAL S.,	Notre experience de la coloscopie peroperatoire. Chirurgie, 97: 409-410; 1971.
DIOT J. MATSUNAGA F.	The clinical value of sigmoidocamera. Proc. Hedrologicum Conlegium, 83-85; 1965.
MATSUNAGA F.	Colonofiberscopy. Proc. IInd World Congress of Gastrointestinal Endoscopy. Rome-Copenhagen 1970.
MATSUNAGA F. et Al.	Fiberscopy for colon diseases and colon biopsy. Rynshogeka, 24: 349; 1969.
MATSUNAGA F., TAJIMA T.	Sigmoidocamera and colonofiberscope. Geka Shinryo (Japanese), Vol. II, pp. 427-433; 1969.
MATSUNAGA F., TAJIMA T.	An outline of colonofiberoscopy. Proceedings of the Second Congress of the International Society of Endoscopy, Copenhagen, 1970.
MATSUNAGA F., TAJIMA T.	Recent advance in colonofiberscopy. Comm. 2e Congres Europ. Endosc. Digest., Paris 1972.
MATSUNAGA F., TAJIMA T., FUKAMI S.	Colonfiberscope. Proc. IInd World Congress of Gastrointestinal Endoscopy, Rome-Copenhagen 1970.
MATSUNAGA F., TAJIMA T., TODA S.	The limits of colonofiberscopy in early diagnosis of malignant and inflammatory changes. Proc. IInd World Congress of Gastrointestinal Endoscopy, Rome-Copenhagen 1970.
MATSUNAGA F., TAJIMA T., TODA S., MAEDA Y., SATO H., MUNAKATA A., HIGUCHI K.	Endoscopie examination of the large bowel. Sogo Rinsho, 19, 325-337; 1970.
MATSUNAGA F., TSUSHIMA H. KUBOTO T.	Photography of the colon. Gastrointest. Endosc. 1: 58-63, 1959.
MEYERS D. S.	Colonoscopy (letter). New Engl. J. Med., 288, 974; 1973.
MYREN J. PETERSEN H.	Sigmoido-colonoscopy in outpatients. Scand. J. Gastroent. 7: suppl. 9, 39; 1972.
MONOD E.	Signes radiologiques et endoscopiques des colites allergiques. Comm. 2° Congres Europeen d'Endoscopie, Paris 1972.
MOORE A. E.	The traction sigmoidoscopy development. Amer. J. Surg., 90, 373, 1955.
MOORE A. E.	Traction sigmoidoscopy. Surg. Cl. N. Amer., 37, 1283; 1957.

MONTORI A., MARTINELLI V., VICECONTE G.	La polipectomia endoscopica del grosso intestino nei soggetti anziani. La Clinica Terapeutica 1974.
MONTORI A., MARTINELLI V., VICECONTE G.	Indicazioni e tecnica della pancolonscopia nei soggetti anziani. Il Prog. Med., 30; 252, 1974.
MONTORI A. VICECONTE G.	La duodenoscopia e la colonscopia peroperatoria. Min. Gastroent., 19 n° 1, 23; 1973.
MORRISSEY J. F.	Gastrointestinal endoscopy. Gastroenterology, 62: 1241-1268; 1972.
MOUTIER F.	Traité de gastroscopie et de pathologie endoscopique de l'estomac. Masson e C. Editeurs - 1935 - Paris.
NAGASAKO K.	Fibercolonoscopy of the terminal ileum. Prof. IInd World Congress of Gastrointestinal Endoscopy, Rome-Copenhagen; 1970.
NAGASAKO K., ENDO M., TAKEMOTO T., KONDO T., KIMURA K.	The insertion of fibercolonoscope into the cecum and the direct observation of the ileocecal valve. Endoscopy, 2: 123-126; 1970.
NAGASAKO K., NAGAI K., SUZUKIH., YAZAWA C., TAKEMOTO T.	Fiberscopic diagnosis of early cancer of the colon. Endoscopy, 4: 1-5; 1972.
NAGASAKO K., TAKEMOTO T.	Fibercolonoscopic diagnosis of the inflammatory diseases. Comm. 2e Congress Europ. Endosc. Digest.; Paris 1972.
NAGASAKO K., TAKEMOTO T.	Endoscopy of the ileo-caecal area. Gastroenterology 65, 403-411; 1973.
NAGASAKO K., TAKEMOTO T.	Fibercolonoscopy without the help of fluoroscopy. Endoscopy, 4, 208-212; 1972.
NAGASAKO K., YAZAWA C., TAKEMOTO T.	Observation of the terminal ileum. Endoscopy, 1/45-51; 1971.
NAGASAKO K., IKEZAWA S. GYO, TAKEMOTO T.	Pra operative, koloskopische diagnose eines nichtoperifischen ulkus in zokum. Dis. Colon Rect. 15, 413; 1972.
NAGY G. S.	Fibrecolonoscopy. Med. J. Aust., 1, 378-382; 1973.
NEDBAL J.	The diagnostic value of fibrosigmoidoscopy. Czech. Cs. Gastroent. Vyziva 27: 467-474; 1973.

NEIGER A.	La sigmoidoscopie a l'àide d'un instrument flexible. Confrontation de la radiologie et de l'endoscopie dans le diagnostic des tumeurs du tube digestif. Comm. 2e Congres Europ. Endosc. Digest.; Paris 1972.
NIWA H.	On photography of the colon and pharynx using. gastrocamera. Gastrointest. Endosc. 2: 77-80, 1960.
NIWA H.	Endoscopy of the colon. Gastroent. Endoscopy, 7: 403; 1965.
NIWA H.	Clinical study of colonfiberscope. Gastroent. Endoscopy, II: 173; 1969.
NIWA H., FUJINO M., UTSUMI Y., KANEKO E., NAKAMURA T., KASUMI A., YOSHIDA A., MATSUMOTO M. YOSHITOSHI Y.	Clinical esperience of colonic fiberscope. Gastroent. Endoscopy, II: 163-173; 1969.
NIWA H., FUJINO M., YOSHITOSHI Y.	Colonic fiberscopy for routine practice. Proc. IInd World Congress of Gastrointestinal Endoscopy; Rome-Copenhagen 1970.
NIWA H., UTSUMI Y., NAKAMURA T., YOSHITOSHI Y.	Endoscopy of the colon. Proc. Ist. Congress of the International Society of Endoscopy; Tokyo, 425; 1966
OI J., TAKEMOTO T., KONDO T., HIRATSUKA H.	Fiberjejunoscopy. Intern. Sympos. Urgent Endoscopy of Digestive and Abdominal Diseases (Abstracts); Praha 1971.
OTTENJANN R.	Colonic polyps and colonoscopic polypectomy. Endoscopy, 4, 212-216; 1972.
OTTENJANN R.	Endoskopie und biopsia des colon. Dtsch. Med. Wschr., 28: 1372; 1968.
OTTENJANN R., DEYHLE P., PAUL F., STADELMANN O.	Die Sondierung des proximalen kolon. Endoscopy, 2: 70-75; 1969.
OTTENJANN R., GRUNER H. J., BARTELHEIMER W.	Fibrocolonscopia parziale ambulatoriale (fibrosigmoidoscopia). Min. Med., 63 n° 93: 5233-5235; 1972. (Trad. da Dtsch. Med. Wschr., 97, 734; 1972).
OVERHOLT B. F.	Clinical experience with the fibersigmoidoscope. Gastroint. Endoscopy, 15, 27; 1968.
OVERHOLT B. F.	Flexible fiberoptic sigmoidoscope. Ca. 19: 81-85, 1969.
OVERHOLT B. F.	Description and experiences with flexible fibersigmoidoscopes. Proceedingis Sixth National Cancer Conference. Philadelphia 1970.

OVERHOLT B. F.	Flexible fiberoptic sigmoidoscopy. Cancer, 28, 123-126; 1971.
OVERHOLT B. F., PALLARD H. M.	Cancer of the colon and rectum. Current procedures for detection and diagnosis. Cancer 20: 445-450; 1967.
PAOLUZZI P.	Total colonoscopy by a "monorail" method. Proc. IInd World Congress of Gastroint. Endoscopy, Rome-Copenhagen 1970.
PAOLUZZI P., PALLONE F., TORSOLI A.	Left and right colon biopsies in the study of proctocolitis. Proc. IInd World Congress of Gastrointestinal Endoscopy, Rome-Copenhagen 1970.
PAPP J. P. HAUBRICH W. S.	Endoscopic removal of colon lipomas. Gastrointestinal Endoscopy 20: 66; 1973.
PAUL F., SEIFERT E., OTTO P.	Results of colonoscopy. Proc. IInd World Congress of Gastrointestinal Endoscopy, Rome-Copenhagen 1970.
PERGOLA P., MARTEAU J., CACHIN M.	La coloscopie dans un cas de pneumatose rectocolique primitive. Comm. 2e Congres Europ. Endosc. Digest., Paris 1972.
PRESTI F., ROSSINI F. P.	Rapporti fra quadri morfologici retto-sigmoidoscopici e colonscopici nella colite ulcero-emorragica. Min. Gastroent., 19 n° 1, 26; 1973.
PROVENZALE L., CAMERADA P., REVIGNAS A.	La coloscopia totale trans-anale mediante una metodica originale. Rass. Med. Sarda, 69: 149; 1966.
PROVENZALE L., REVIGNAS A.	Metodica originale di esplorazione strumentale trans-anale polivalente del colon interno. Rass. Med. Sarda, 89: 131; 1966.
PROVENZALE L., REVIGNAS A.	Transanale koloskopie mittels fiberoptik. Bericht einer neuen methode zur untersuchung des gesamten dickdarms. Fortschr. Endosk., 1, 167; 1968.
PROVENZALE L., REVIGNAS A.	Ergebnisse und moglichkeiten der fiberkolonoskopie. in: R. Ottenjann (ed), Fortschritte der Endoskopie, B.D.I. Schattauer, Stuttgart 1969.
PROVENZALE L., PISANO M., REVIGNAS A.	Moderne metodiche di studio e rilievi fisiopatologici nelle malattie del colon di interesse chirurgico. Relazione LXVIII Congresso Società Italiana Chirurgia, Roma 1966.
PROVENZALE L., REVIGNAS A.	Ergebnisse und moglichkeiten der fiberkolonoskopie. Fortschr. Endosk., 2, 213; 1969.

PROVENZALE L., REVIGNAS A.	An original method for guided intubation of the colon. (Technique and applications for different types of transanal exmination of the large bowel). Gastroint. Endoscopy, 16, 11; 1969.
PROVENZALE L., REVIGNAS A.	Colonic electrical activity recording and total fiberoptic colonoscopy. Amer. J. Gastroent., 56, 137-148; 1971.
REGINS H., SHINYA H.	The explosive potential of colonic gas during colonscopic elettrosurgical polypectomy. Surgery, Ginecology and obstetrics, 138: 554-556; 1974.
REIFFERSCHEID M.	Entstehung und verhutung des lokalen rezidivs nach resektion des kolonkarzinoms. Akt. Chir., 1: 21; 1966.
REISS H., RYE K.	Roslegly naczyniak jelita grubego i odbytnicy o mieszaney budowie. Pol. Przegl. Chir., 43: 115-118; 1971.
REVIGNAS A., DI MARTINO L., PROVENZALE L.	Total non-operative colonoscopy by means of fiberoptic instruments (up to date techniques and results). Proc. IInd World Congress of Gastrointestinal Endoscopy, Rome-Copenhagen 1970.
ROGERS B. H. G.	The safety of carbon dioxide insufflation during colonoscopy. Gastrointestinal Endoscopy, 20: 115-117; 1974.
ROSENBERG J. C., DI DIO L. J. A.	Anatomic and clinical aspect of the junction of the ileum with the large intestine. Dis. Col. & Rect., vol. 13, n° 3, MayJune 1970.
ROSSINI F. P.	La colonscopia d'urgenza nelle enterorragie. Acta Endoscopica et Radiocinématographica - Tome III n° 5 - pagg. 131-132; nov. 1973.
ROSSINI F. P.	Coloscopy and endoscopic polypectomy in the prevention of cancer of the large bowel. Pan Minerva Medica—in press.
ROSSINI F. P., ROATTA L., PRESTI BOIDO C., FERRARI A.	La colonscopia nel controllo a distanza delle anastomosi chirurgiche del grosso intestino. V Congresso Naz. Soc. Endoscopia Digestiva, Venezia, settembre 1973.
ROSSINI F. P., BONARDI L.	Diagnosi colonscopica nei confronti delle altre metodiche di diagnosi precose dei tumori del colon. Min. Gastroent., 19 n° 1, 24 1973.
ROSSINI F. P. et Al.	Possibilità e limiti della polipectomia endoscopica - Casistica. Min. Gastroent. in press.
SAKAI Y.	Further progress in coloscopy. Gastrointestinal Endoscopy, 20: 143; 1974.

SAKAI Y., ASHIZAWA S.	Fiberscopic examination of the rectum and sigmoid colon. Gastroent. Jap., 5: 281; 1970.
SAKAI Y.	The technic of colonofiberscopy. Dis. Colon Rect., 15, 403-412; 1972.
SALMON P. R., BRANCH R. A., COLLINS C., ESPINNER H., READ A. E.	Clinical evaluation of fiberoptic sigmoidoscopy employing the olympus CF—SB colonoscope. Gut, 51, 729-735; 1971.
SALMON P. R., TEAGUE R., READ A. E.	Fibre—optic examination of the large bowel: clinical results in 200 cases. Comm. 2e Congres. Europ. Endosc. Digest.; Paris 1972.
SCHULKE K., BECKER H.	Intraoperative endoskopische tumordiagnostik des dichdarms. Hedrologicum Conlegium; Budapest 1970.
SCHULKE K., UNGEHEUER E.	Die coloskopie, eine methode der intraoperativen endoskopischen untersuchung des kolons. Langenbecks Arch. Klin. Chir., 322, 780; 1968.
SEGAL S., DIOT J., MAFFIOLI C.	Confrontation radio-endoscopique dans l'étude du colon terminal. Endoscopy, 3: 183-187; 1971.
SEGAL S., DIOT J., MAFFIOLI C., SEGAL A., GEOFFROY H.	La coloscopie trans-anale. Maroc. Med., 54: 283-299; 1971.
SEGAL S., DIOT J., SEGAL A.	Les difficultés de la coloscopie de routine. Acta gastro-ent. belg., vol. 37, 140-143; 1974.
SEIFERT E., PAUL F., OTTO P.	Value of colonscopy. Proc. IInd World Congress of Gastrointestinal Endoscopy, Rome-Copenhagen 1970.
SEIFERT E.	A new flexible forceps for the removal of polyps from the gastrointestinal tract after successful endoscopic polipectomy. Endoscopy, 4, 226-227; 1972.
SEIFERT E.	Endoskopische polypektomie in verdauonestrakt. KongreB der Deutschen Gesellschaft fur Endoskopie Erlangen 1973.
SHEARMAN D. J. C.	Colonoscopy in ulcerative colitis. Scand. J. Gastroent. 8: 289-291; 1973.
SHINYA H.	Removal of colonic polyps by fiberoptic colonoscopy. New Engl. J. Med., 288, 329-332; 1973.
SHINYA H., WOLFF W. I.	Fiberoptic endoscopy of the entire small and large intestine. Comm. 2e Congress Europ. Endosc. Digest., Paris 1972.

SHINYA H., WOLFF W.	Therapeutic applications of colonfiberoscopy: polypectomy via the colonfiberscope. Gastroenterology (abstr.) 60: 830; 1971.
SHINYA H., WOLFF W.	Therapeutic applications of colonfiberoscopy. Polypectomy via the colonoscope. Comm. 2° Congres Europeen d'Endoscopie—Paris 1972.
SIMON C.	Apport de l'endoscopie dans les polypes et polyposes recto-coliques de l'adulte. Ann. Chir., 25: 597-601; 1971.
SIVAK M. V. Jr., SULLIVAN B. H. Jr., WEAKLEY F. L., RANKIN G. B.	Colonoscopic polypectomy. Digestive Diseases, vol. 19, n° 4 (april 1974).
SOULLARD J., POTET F., LANDRIEU P.	Les polyposes juveniles diffuses. Arch. Franc. Mal. Appar. Dig., 61: 391-394; 1972.
SPIRO H. M.	Clinical gastroenterology. Vol. II - Mac Millan Company - London, 1970.
SUGARBAKER P. H., VINEYARD G. C.	Snare polypectomy with the fiberoptic colonscope. Surgery, Gynecology and obstetrics, 138: 581-3; 1974.
SWINTON N. W.	Colotomy and coloscopy. Surg. Clin. North Am., 40: 727-731; 1960.
SWINTON N. W., WEAKLEY F. L.	Complications of colotomy and colonoscopy. Dis. Colon Rect., 6, 50-53; 1963.
TEAGUE R. H., SALMON P. R., READ A. E.	Fiberoptic examination of the colon: a review of 255 cases. Gut, 14, 139-142; 1973.
TESTAS P., CHAVY A., LIGUORY C., NAUDIN G.	L'exploration du colon par fibrocoloscopie et la polypectomie endoscopique. Chirurgie, 99 pp. 485-488; 1973.
TODA S., TAJIMA T., MATSUNAGA F.	Endoscopic and histologic studies of the chronic ulcerative colitis. Comm. 2e Congres Europ. Endosc. Digest., Paris 1972.
TORSOLI A., ARULLANI P., CASALE C.	An application of transintestinal intubation to the colon. Gut, 8: 192; 1967.
TORSOLI A., ARULLANI P., PAOLUZZI P.	Transintestinale sondierung als leitmethode fur kolombiopsie, endoskopie und intraluminale studien. Fortschritte der Endoskopie, B.D.I. Schattauer Stuttgart 1969.
TURELL R.	Fiberoptic coloscope and sigmoidoscope. Preliminary report. Amer. J. Surg., 105, 133-136; 1963.
TURELL R.	Fiberoptic sigmoidoscopes. Am. J. Surg. 113: 305-307; 1967.

TUTTLE J. P.	A treatise on diseases of the anus rectum and pelvic colon. Appleton & Co., New York, 1905.
UMEDA K. et Al.	A very Rare Case of Leiomyoma of the Transverse colon. Stomach Intestine 8: 1471; 1973.
UKHVIDOVA J. M., ZINOVIEV O. I., TIKHONOVA A. F.	Colonfiberscopy in colon diseases diagnostics. Comm. 2e Congres Europ. Endosc. Digest., Paris 1972.
YAMAGATA S. et Al.	Clinical experience of fibercolonoscope type VI. Gastroent. Endoscopy, II: 219; 1969.
YAMAGUCHI T., MATSUNAGA F.	Studies on photographing the colonic mucosa. Proc. Ist Congress of the International Society of Endoscopy, pp. 398-403; 1966.
YUKAWA E.	Examination of upper colon with combined use of x-ray and colonofiberscope. Abstracts of 4th World Congress of Gastroenterology, Copenhagen 1970.
WAGGENER H. U., FERGUSON J. A.	Experience with total colonscopy in 50 patients. J. Michigan M. Soc., 58, 619; 1960.
WARWICK R. R. G., SUMERLING M. D., GILMOUR H. M., SHEARMAN D. J. C.	Colonoscopy and double contrast barium enema examination in chronic ulcerative colitis. Amer. J. Roentgenol., 117, 292-296; 1973.
WATANABE H.	Personliche mitteilung. Juni 1970.
WATANABE H., NARASAKA T., UEZU J., MIURA K., YAMAGATA S.	An improved fibercolonoscope and its application to the colonic diseases. Proc. IInd World Congress of Gastrointestinal Endoscopy, Rome-Copenhagen 1970.
WAYE J. D.	Coloscopy. Surg. Clin. N. Amer., 52, n° 4; 1972.
WILLIAMS C. B.	Diathermy-biopsy: a technique for the endoscopic management of small polyps. Endoscopy, 5, 215-218; 1973.
WILLIAMS C. B.	Coloscopy: a critical comment. Acta gastro-ent. belg., Vol. 37, 129-137; 1974.
WILLIAMS C. B., LANE R., SAKAI T., HANWELL A.	Air pressure: a hazard during colonoscopy. Lancet II, 729; 1973.
WILLIAMS C. B., LANE R. H., SAKAI Y.	Colonoscopy: an air-pressure hazard. Lancet, September 29; 1973.
WILLIAMS C. B., TEAGUE R. M.	Colonoscopy progress report. Gut, 14, 990-1003; 1973.

WILLIAMS C. B., MUTO T.	Examination of the whole colon with the fibreoptic colonoscope. Brit. Med. J., 3: 278-281; 1972.
WILLIAMS C. B., MUTO T., UTTER	Removal of polyps with fibreoptic colonoscope: a new approach to colonic polypectomy. Brit. Med. J., 1, 451-452; 1973.
WOLFF W. I.	Editorial - Colonoscopy and endoscopic polypectomy. N. Y. State J. Med. 73: 641-642; 1973.
WOLFF W. I., SHINYA H.	Colonofiberoscopy. JAMA, 217: 1509-1512; 1971.
WOLFF W. I., SHINYA H.	Colonofiberscopic management of colonic polyps. Dis. Colon Rectum 16: 87-93; 1973.
WOLFF W. I., SHINYA H.	Colonofiberoscopy—Diagnostic modality and therapeutic application. Bull. Soc. Int. Chir. 5: 525-529; 1971.
WOLFF W. I., SHINYA H.	Earlier diagnosis of cancer of the colon through colonic endoscopy (Colonoscopy). Cancer 34: 912-931; 1974.
WOLFF W. I., SHINYA H., EFFEN A., ZAKTAY S. Z.	Colonofiberscopy. A new and valuable diagnostic modality. Amer. J. Surg. 123: 180-184; 1972.
WOLFF W. I., SHINYA H.	Polypectomy via the fiberoptic colonoscope. N. Engl. J. Med., 288, 329-332; 1973.
WOLFF W. I., SHINYA H.	A new approach to colonic polyps. Ann. Surg. 178: 367-375; 1973.
EIMER H., SIMKIN A.	A frictionless method of entering the colon. Proc. IInd World Congress of Gastrointestinal Endoscopy. Rome-Copenhagen 1970.

inito di stampare
el mese di ottobre 1975
resso l'Istituto Grafico Litostampa
i Gorle (Bergamo)
er conto di Piccin Editore, Padova